CW00363725

OVID'S *METAMORPHOSES* AND THE ENVIRONMENTAL IMAGINATION

ANCIENT ENVIRONMENTS

Series Editors: Anna Collar, Esther Eidinow and Katharina Lorenz

The Ancient Environments series explores the worlds of living and non-living things, examining how they have shaped, and been shaped by, ancient human societies and cultures. Ranging across the Mediterranean from 3500 BCE to 750 CE, and grounded in case studies and relevant evidence, its volumes use interdisciplinary theories and methods to investigate ancient ecological experiences and illuminate the development and reception of environmental concepts. The series provides a deeper understanding of how and why, over time and place, people have understood and lived in their environments. Through this approach, we can reflect on our responses to contemporary ecological challenges.

Also available in the series

MOUNTAIN DIALOGUES FROM ANTIQUITY TO MODERNITY
edited by Dawn Hollis and Jason König

SEAFARING AND MOBILITY IN THE LATE ANTIQUE MEDITERRANEAN
edited by Antti Lampinen and Emilia Mataix Ferrándiz

THE SPIRITED HORSE: EQUID–HUMAN RELATIONS IN THE BRONZE AGE NEAR EAST
by Laerke Recht

OVID'S *METAMORPHOSES* AND THE ENVIRONMENTAL IMAGINATION

Edited by Francesca Martelli and Giulia Sissa

BLOOMSBURY ACADEMIC
LONDON • NEW YORK • OXFORD • NEW DELHI • SYDNEY

BLOOMSBURY ACADEMIC
Bloomsbury Publishing Plc
50 Bedford Square, London, WC1B 3DP, UK
1385 Broadway, New York, NY 10018, USA
29 Earlsfort Terrace, Dublin 2, Ireland

BLOOMSBURY, BLOOMSBURY ACADEMIC and the Diana logo are trademarks of Bloomsbury Publishing Plc

First published in Great Britain 2023

A catalogue record for this book is available from the British Library.

Library of Congress Cataloging-in-Publication Data
Names: Martelli, Francesca, 1978- editor. | Sissa, Giulia, 1954- editor.
Title: Ovid's Metamorphoses and the environmental imagination / Francesca Martelli and Giulia Sissa.
Other titles: Ancient environments.
Description: New York : Bloomsbury Academic, 2023. | Series: Ancient environments | Includes bibliographical references and index.
Identifiers: LCCN 2022049371 | ISBN 9781350268944 (hardback) | ISBN 9781350268982 (paperback) | ISBN 9781350268951 (ebook) | ISBN 9781350268968 (epub)
Subjects: LCSH: Ovid, 43 B.C.-17 A.D. or 18 A.D. Metamorphoses. | Environmentalism in literature.
Classification: LCC PA6519.M9 O97 2023 | DDC 871/.01—dc23/eng/20221110
LC record available at https://lccn.loc.gov/2022049371

ISBN: HB: 978-1-3502-6894-4
ePDF: 978-1-3502-6895-1
eBook: 978-1-3502-6896-8

Series: Ancient Environments

Typeset by RefineCatch Limited, Bungay, Suffolk

To find out more about our authors and books visit www.bloomsbury.com and sign up for our newsletters.

CONTENTS

ACKNOWLEDGEMENTS

We would like to thank Massimo Ciavolella, former Director of the Center for Medieval and Renaissance Studies and Zrinka Stahuljak, Director of the CMRS Center for Early Global Studies at UCLA for supporting the international conference that provided the original spur for this book ('Metamorphosis and the Environmental Imagination, from Ovid to Shakespeare', October 11–12, 2019). We also thank Anthony Pagden for his collegial help in preparing this collection of essays.

John Shoptaw's poem *Whoa!* was first published in 2019 by *Arion* 27.1: 1–20.

Shane Butler's essay in this volume is a revised and expanded version of his essay 'Animal Listening', originally published in 2021 by *Journal of International Voice Studies* 6.1: 27–38.

CONTRIBUTORS

Shane Butler is the Hall Professor in the Humanities and Professor of Classics at Johns Hopkins University. With primary interests in aesthetics and queer theory, he has published widely on classical literature and its reception, Renaissance humanism, the history of sensation, the phenomenology of reading, and the history of sexuality. His most recent books are *The Ancient Phonograph* (2015), *Sound and the Ancient Senses*, co-edited with Sarah Nooter (2019), and *The Passions of John Addington Symonds* (2022).

Sandra Fluhrer is Assistant Professor of Comparative and German Literature at the University of Erlangen-Nuremberg, Germany. Her research focuses on theories and practices of aesthetic experience, forms of theatricality, and the relationship between literature, mythology, and the political. Her new book on the aesthetics and politics of metamorphosis in European literature and political philosophy is soon to be published.

Marco Formisano is Professor of Latin Literature at Ghent University (Belgium). He has published extensively on late antique literature as well as on early Christian martyr acts (in particular Perpetua and Polycarp). Other research interests include ancient technical texts, Ovid's *Metamorphoses*, and Classical reception in film and literature. He is editor of the series "*sera tela*. Studies in Late Antique Literature and Its Reception" (Bloomsbury) and "The Library of the Other Antiquity" (Universitätsverlag Winter, Heidelberg) and he is the Principal Investigator of the research project "Coming After. Late Antique Ecopoetics", funded by FWO, Research Foundation Flanders.

Emily Gowers is Professor of Latin Literature at the University of Cambridge and a Fellow of St John's College. She has written on many aspects of the natural, cultural, and material worlds of ancient Rome, including food, insects, owls, travel, gardens, and sewers. This is her third article on trees. She edited *Horace, Satires Book I* for the Cambridge Green and Yellow series and in 2014–16 she held a Leverhulme Major Fellowship to write a book on Maecenas (forthcoming). In Spring 2022 she was Sather Professor at the University of California at Berkeley, where she gave lectures on "The Small Stuff of Roman Antiquity".

Miranda Griffin is Assistant Professor of Medieval French Literature at the University of Cambridge, where she is a Fellow of Murray Edwards College. She is the author of two monographs: *The Object and the Cause in the Vulgate Cycle* (2005) and *Transforming Tales: Rewriting Metamorphosis in Medieval French Literature* (2015). With Jane Gilbert, she is the co-editor of *The Futures of Medieval French Literature: Essays in Honour of Sarah Kay*, (2021). Miranda has published articles on a wide variety of medieval French literary culture, and is currently working on a book about imagined landscapes in medieval French literature.

Julia Reinhard Lupton is Distinguished Professor of English at the University of California, Irvine, where she has taught since 1989. She is the author or co-author of five books on Shakespeare, including *Shakespeare Dwelling: Designs for the Theater of Life* (2018). She is a former Guggenheim Fellow.

Francesca Martelli is Associate Professor of Classics at the University of California, Los Angeles. She is the author of two books on Ovid (*Ovid's Revisions*, 2013; and *Ovid*, 2020), and of a forthcoming book on Cicero's letter collections. Her current research interests take in contemporary ecocritique (and its application to ancient Greek and Roman thought), as well as the meanings and affective valences of political loss in the texts of the late Roman Republic and their resonance with our current historical moment.

John Shoptaw was raised in the drained Mississippi River wetlands of the Missouri Bootheel. He lives now in the San Francisco Bay Area and teaches poetry and ecopoetry at UC Berkeley. He is the author of *On the Outside Looking Out: John Ashbery's Poetry* (1996). He wrote the libretto for an Eric Sawyer opera *Our American Cousin* (2008). His collection of poems, *Times Beach* (2016), won the Northern California Book Award in Poetry. First published in *Arion* (2019), his poem "Whoa!" will be included in his new poetry collection, *Near-Earth Object*.

Giulia Sissa is Distinguished Professor in the Departments of Political Science, Classics and Comparative Literature at UCLA. While anchoring her research to the societies and the cultures of the Greek and Roman world, Giulia connects the study of the past to moments of reception, modern recontextualizations, and significant resonances in the contemporary world. Her publications include *Greek Virginity* (2000) *The Daily Life of the Greek Gods*, with M. Detienne, (2000); *Le Plaisir et le Mal. Philosophie de la drogue* (1997); *L'âme est un corps de femme* (2000); *Sex and Sensuality in the Ancient World* (2008); *Utopia* 1516–2016. *More's Eccentric Essay and its Activist Aftermath*, co-edited with Han van Ruler (2017); *Jealousy. A Forbidden Passion* (2017); *Le Pouvoir des femmes. Un défi pour la démocratie* (2021); *A Cultural History of Ideas in Classical Antiquity*, co-edited with Clifford Ando (Bloomsbury, 2023).

Diana Spencer is Professor of Classics at the University of Birmingham, UK. Most recently, she is the author of *Language and Authority in De Lingua Latina: Varro's Guide to Being Roman* (2019), *Roman Landscape: Culture and Identity* (2010), co-editor of *The Sites of Rome: Time, Space, Memory* (2007), and has written on a wide range of Latin authors. She has ongoing research interests in the relationship between urban and rural spaces and intellectual culture in ancient Rome, and is currently working on the role of metals as urban landmarks and placemakers.

Claudia Zatta is the author of *Interconnectedness. The Living World of the Early Greek Philosophers* (2017, second revised edition, 2019) and of *Aristotle and the Animals. The Logos of Life Itself* (2022), in addition to numerous articles that have appeared in Europe and the United States. She is a researcher in the department of philosophy at the University of Milan.

SERIES PREFACE

While our intention in writing this preface was to provide a neutral introduction that could stand for the whole series, recent events are too dramatic and relevant to ignore. As we launch the series, and write this text, we are (hopefully) emerging from the ravages of the 2020 Covid-19 pandemic. Along with the climate crisis, this experience has increased awareness of human reliance and impact on the environments we occupy, dramatically emphasized human inability to control nature, and reinforced perceptions that the environment is the most pressing political and social issue of our time. It confirms our belief that the time is right to situate our current (abnormal?) relationship with nature within an examination of human interactions with the environment over the *longue durée* – a belief that has given rise to this series.

Ancient Environments sets out to explore (from a variety of perspectives) different constructions of the 'environment' and understandings of humankind's place within it, across and around the Mediterranean from 3500 BCE–750 CE. By 'environment' we mean the worlds of living and non-living things in which human societies and cultures exist and with which they interact. The series focuses on the *co-construction* of humans and the natural world. It examines not only human-led interactions with the environment (e.g., the implications of trade or diet), but also those that foreground earth systems and specific environmental phenomena; it investigates both physical entities and events and ancient, imagined environments and alternate realities. The initial and primary focus of this series is the ancient world, but by explicitly exploring, evaluating and contextualizing past human societies and cultures in dialogue with their environments, it also aims to illuminate the development and reception of environmental ideas and concepts, and to provoke a deeper understanding of more long-term and widespread environmental dynamics.

The geographical remit of this series includes not only the cultures of the Mediterranean and Near East, but also those of southern Europe, North Africa including Egypt, northern Europe, the Balkans and the shores of the Black Sea. We believe that encompassing this broader geographical extent supports a more dynamic, cross-disciplinary and comparative approach – enabling the series to transcend traditional boundaries in scholarship. Its temporal range is also far-reaching: it begins with the Neolithic (a dynamic date range, depending on location in the Near East/Europe) because it marks a distinct change in the ways in which human beings interacted with their environment. We have chosen *c.* 750 CE as our end date because it captures the broadest understanding of the end of Late Antiquity in the Central Mediterranean area, marking the rise of the Carolingians in the West, and the fall of the Ummayyad Caliphate in the East.

Our series coincides with, and is inspired by, a particular focus on 'the environmental turn' in studies of the ancient world, as well as across humanities more generally. This focus is currently provoking a reassessment of approaches that have tended to focus solely on people and their actions, prompting scholars to reflect instead (or alongside) on the key role of the environments in which their historical subjects lived, and which shaped and were shaped by them. By extending beyond the chronological and geographical boundaries that often define – and limit – understanding of the meaning of 'antiquity', we intend that this series should encourage and enable broader participation from within and beyond relevant academic disciplines. This series will, we hope, not only advance the investigation of ancient ecological experiences, but also stimulate reflection on responses to contemporary ecological challenges.

The editors would like to express heartfelt thanks to Alice Wright at Bloomsbury Press who first conceived the idea and suggested it to Esther, and who has done so much to develop it, and to Georgina Leighton, in particular for her work in launching the series. We are extremely grateful to the members of the Series Board, who have provided such wonderful encouragement and support, and to our authors (current and future) who have entrusted their work to this 'home'. We have chosen the 'Mistress of Animals' or *Potnia Theron*, a figure found in Near Eastern, Minoan, Mycenean, Greek and Etruscan art over thousands of years, as the motif for the series.

Anna Collar
Esther Eidinow
Katharina Lorenz

INTRODUCTION

Francesca Martelli and Giulia Sissa[1]

Bodies. Change. Deep ecological time. The themes that Ovid summarily announces in his proem to the *Metamorphoses* speak readily to the preoccupations of many discourses currently being used to describe the place of the human within the systems of planetary life in which s/he is enmeshed. The corporeality of the changes that he describes situates the poem within a materialist tradition that stretches from the Presocratics to modern science.[2] That tradition has only gained in urgency as the transformative impact that humans have had on the material biosphere has become more visible, not least in its effect on the human body. If Virgil inaugurates an ecocritical tradition that laments the former process, Ovid's poem stands at the head of a tradition that takes a slightly different line.[3] It too denounces the destruction posed by the human disruption of the wider ecosphere, but it frequently does so by demonstrating the dangerous consequences that such intervention holds for human beings: the poem's persistent take on metamorphosis, which describes the transformation of humans into other life forms (rather than the other way around), speaks presciently to the capacity for ecological change to be felt by, and imprinted on, the human. Anthropos, the sovereign subject that Enlightenment Europeans conjured as the legacy of Classical Greece, is being unmade in disciplines as diverse as philosophy, cultural anthropology, evolutionary biology and atmospheric chemistry, thanks in part to developments in cultural and scientific understanding, and in part to a raised awareness of the harm that the assumed sovereignty of the human has imposed on planetary life.[4] Ovid's *Metamorphoses* reminds us, more insistently than any other single text from ancient Greece or Rome, that that unmaking is already an important story within antiquity, and that, as far as Anthropos is concerned, we should take the Enlightenment's projections onto the Classical past as partial, at the very least.

The purpose of this volume is to track some of the ecological meanings that are implicit in Ovid's stories of that unmaking, and to frame these meanings in light of various lines of contemporary ecocritique.[5] While the Latin language has no specific word for either the environment or ecology (and its cognates), the *Metamorphoses* illustrates a foundational premise of much modern environmental and/or ecological thought in its display of the highly porous relationship between 'nature' and 'culture'. In this, it exemplifies Bruno Latour's reason for advocating a return to the thought structures and patterns of the 'premodern' world: namely, their ability to straddle and hybridize disparate discourses (scientific, political and even poetic) in order to allow humans to conceptualize the interconnections between the human and the physical world. This is one of the chief imperatives of environmental criticism, in all its many stripes, and it is one to which Ovid's poem repeatedly answers.[6] In this Introduction, we will offer a brief sketch of the ecological discourses with which our papers engage, tracing the way in which these discourses relate to one another, signposting where our contributors advance this narrative, and demonstrating (briefly) their bearing for Ovid's poem (and vice versa). Our discussion tracks in and out of the philosophical traditions that have broached various ecological concerns – about ethics, materialism, subjects and objects – and picks up those discourses that take over where philosophy's answers to those concerns fail to address their concrete manifestation in the earth's actual life systems and actors. Within the sprawling field of contemporary ecocriticism, philosophy's abstractions frequently yield to the observational techniques of various forms of ethnography and to anthropological theories that are based on concrete processes of observational attention. As curious students of this field, we attempt in this Introduction to keep pace with the way in which these disparate discourses pick up from, fall short of and overlap with one another, and to demonstrate where and how Ovid's poem is positioned to intervene in them. The *Metamorphoses* emerges from our study as a text riven by disparate ecological visions: one in which systems of life are sustained in symbiotic balance, and another in which the differences between life forms collapse in on each other, threatening (and producing) untold forms of destruction. To complicate this double strategy even more, the poem alerts its readers to the intrinsic hazards of metamorphosis itself. Non-human incarnations of human subjectivity create confusion, violence and even cannibalism. In presenting all these visions, the poem mirrors both the aspirations and the fears of contemporary ecological thought.

Anthropology, animism and metamorphosis

All things are in flux (*cuncta fluunt*).[7] This is Pythagoras speaking in Book 15 of the *Metamorphoses*, but this general premise runs throughout the myriad narratives that make up the poem. The first two lines announce a storytelling of *nova corpora*, the new bodies into which multiple beings, more precisely human beings, are to be converted.[8] The poem will detail their becoming, namely their coming to be otherwise – in a variety of non-anthropomorphic re-embodiments.

Ovid creates a possible world, in which new bodies can be generated all the time. And this because the gods transform themselves – think of Zeus morphing into a cow, an eagle, a shower of rain – and transform human beings, as in the rest of the poem. A possible world is not simply a fictional world. In the wake of modal semantics, a *possible* world may well be fantastic and unrealistic, but is apprehended in its internal coherence. A variety of necessary and conceivable consequences – and not merely a sequence of *de facto* invented situations – unfolds from postulates, premises and dispositions. Once a text sets a principle – for instance, by announcing: 'I will tell of shapes changed into new bodies' – we are invited to ask questions as to what is bound to happen, what is admissible, probable, inevitable or off-limits. We are expected to ask these questions, to probe the logic of the events. The text may well surprise us, but it must abide by the law it has laid down.[9] We also may consider how far what is imaginable for figures within a poem may be feasible for us too. This is how we cooperate to make sense of what we read.[10] Characters are such that we can anticipate or at least try to guess what might happen to them. What may well be impossible for us, metamorphosis, is possible for them.

The *Metamorphoses* is a thought experiment in which this logical modality – the possibility of the impossible – is incessantly explored: the fifteen books of the poem contain more than 200 tales, making the text an open-ended meditation about the very possibility of metamorphosis. Simply by repetition, this deliberately innovative epic poem confirms that the sudden disfigurement of a human being *can* happen – and may happen now at once, at any time. Seriality generates probability. In the array of cosmological and non-cosmological, reversible and irreversible changes, this is what brings together all the episodes. Transformation occurs as an accident, but, because it does so over and over again, verse after verse, book after book, it becomes a systemic occurrence. Rather than emphasizing the randomness of individual metamorphoses, therefore, we should comprehend the logical structure of this infinitely expanding universe. This is an environment where shapes may *always* shift. As Pythagoras puts it, the ecosphere undergoes a continuous renovation.[11]

Five premises ground such a reading of the poem. First, the flux is active, lively, productive. The entire poem is meant to offer an ontological understanding of the environment, as opposed to a merely 'ontic' description of what is already there. Beings are apprehended in the contingent event of their coming to be in the world, right now. They may well not exist at all; they may never have come into existence. But there they are! Secondly, the prospect of metamorphosis defines this textual world. The frequency of so many alterations of the anthropomorphic shape determines the likelihood of change as a permanent ontological risk. Thirdly, such a risk is treacherous.

Metamorphosis may and can occur any minute, or might have occurred already, unbeknownst to ourselves. Uncertainty can elicit wonder, to be sure, but also loss, horror and fear. One disquieting chance, that of a corporeal makeover, sets the stage for the entire narrative: by modifying pre-existent, adult, human individuals who were already there, metamorphosis extends indefinitely the potential to generate new beings. The poem, therefore, is a meditation not only on the probability of metamorphosis itself, but also on its epistemic consequences. Change is incomplete. These newly modified individuals are hybrids. In the interplay of *ipse/idem* or *ipsa/eadem*, an old subjectivity, an autobiographical memory and a familiar intentionality remain alive while a differently-able anatomy now has come into existence. It is much more than a similarity between old and new, it is an invisible presence.[12] Humans have changed beyond recognition. If this is possible, then what else could possibly happen? Who knows?[13] In any event, *we* are in danger. Fourthly, the danger affects us as members of humankind, but also as spectators of the poem's possible world. In the book, all humanlike beings are the potential victims of alteration. How about us, here and now? Are we too dwelling in a fluid universe? Finally, our last premise is that change itself changes. The temporality of metamorphic becoming is both discontinuous and slow. Freeze frame: the stone who/that is Niobe, suddenly petrified and still weeping, is going to be there for some time. It is there. She is there.

By creating a possible world in which a human form can always yield to a new body, the poem offers a narrative reflection on what it means to be 'specifically' human. In the tradition of taxonomic thought inaugurated by Aristotle, the *anthropos* belongs in a stable system of analogies and distinctions vis-à-vis all other living beings. Metamorphic thinking dislodges humanity from its unique place. That the beings undergoing metamorphosis should be our fellow humans – except when Pyrrha and Deucalion convert stones into women and men, which restart the process of anthropogenesis – can hardly be a coincidence. In her chapter, Claudia Zatta reminds us that these ancestral people have themselves a nonhuman provenance from mineral parcels of the earth. Yet, they proliferate as our own human progenitors. To be human in this metamorphic environment means, on the one hand, to be neatly distinct from an herbivorous quadruped, from an evergreen bush and even from a new block of granite; on the other, it means to be exposed to the fluidification of that distinction, to the possibility of acquiring a new version of who/what we are. Metamorphosis, therefore, is all about us. It is about how a human existence can slide, glide, melt, fuse, dissolve or harden into a more complicated existence – still there, but now looking like a laurel tree, and moving, signifying intent as only a laurel tree can possibly do. It is about our own morphological vulnerability.

This is why Ovid's *Metamorphoses* is a piece of anthropology. It starts as a narrative ready-made for the structural analysis practised by Claude Lévi-Strauss.[14] It offers a classification of beings – plants, minerals, stars, rivers, springs, nonhuman and human animals – which have emerged from an indistinct origin. The universe expands through a progressive differentiation. These beings multiply and proliferate in different spaces, being equipped with dissimilar yet matching features. What is hair for a woman is leaves for a tree. But the poem itself sets analogy into narrative motion. The poet selects relevant qualities available in his culture, and combines them in a sequence. Paradigmatic possibilities are actualized in syntagmatic emplotments. It is a twofold form of relationality. For, as Emily Gowers emphasizes in her paper on the similarities between trees and humans, the dissimilarities are just as important: a young woman is *not* a vegetable. Her hair, arms, face and swift feet are different from leaves, branches and roots, yet the parts of these two bodies stand in a rapport of equivalence. The roots of a bush correspond to the lower limbs of a human being. The poem invites us to observe this analogy.

But it does much more. This particular person, Daphne, is now running from an unwelcome suitor, the young god, Apollo, and is in the process of becoming a particular plant, the laurel tree, her mane now trembling in the wind under the species of delicate leaves. Analogy in narrative motion becomes metamorphosis. This *can* happen, because this poetic world, as we have said, is defined by such possibility, stipulated at the beginning – and such possibility is what matters. It is not the case that all women are also trees, or that all trees are females or that all females are prone to become arboreal vegetation.[15] Quite the opposite. The cosmos is neatly set out. Rather, it is the case that the specificity of a human body is *compatible* with an individual, eventful, ever-probable disfiguration or refiguration. This twofold feature – paradigmatic taxonomy and potential change in the course of a narration – is systematic. Structural analysis is indeed attentive to intrigues, twists and turns in what we call 'myths', but it looks for patterns. Its purpose is to examine how the ideas that belong to the ethnographic context, namely the imagination and knowledge shared in a given society about the environment, have been placed and interconnected in a story-line. The structure of a myth is made up of mythemes, namely of narrative segments that may either recur or reverse each other, in parallelistic arrangements. The essential objective of a structural analysis is to identify how a myth recounts the transformation of an initial state of affairs, by repeating, permutating, inverting topical moments. More ambitiously, such an analysis tracks down variations from one myth to another.

Now, a thought-provoking new trend has emerged, which is shifting the very project of contemporary anthropology. A pupil of Lévi-Strauss, Philippe Descola, has argued that different societies live within different ontologies.[16] One of them is animism. The world is one ecosphere, in which all beings may well inhabit the most disparate bodies, but nevertheless share a common subjectivity. The same core of conscience, agency and selfhood is susceptible of belonging to anything that exists, no matter how that thing has come to be – growing spontaneously, being carefully cultivated, being crafted by an artisan, being born of parents, or being otherwise transformed. Because variety is so vast, the people who live in an animist society may occasionally fail to perceive their shared identity. But the commonality is there, and it is modelled on human capacities. It is the human soul that can be, or move, everywhere. An animist mythology, therefore, is keen on narrating precisely how seemingly nonhuman animals, plants or minerals are animated like humans, or even derive from humans through a morphological transformation. These ontologies project human animation onto all other beings, and conceive cosmogony as 'anthropogenesis', meant as a descent from human beings. The *anthropoi* have no privileges, paradoxically, because they are everywhere, and engender everything.

In this ontological perspective, nature is inseparable from what anthropologists used to call 'culture', namely the social, technical and symbolic creations of a society. Descola's ground-breaking book is programmatically entitled *Beyond Nature and Culture* because, unlike Western, modern ontologies, animism presupposes the absence of such a neat distinction. Rather, the same *anima* or *animus* moves, travels and crosses boundaries, being able to animate any possible part and parcel of the cosmos. Shamanism and, unsurprisingly, metamorphosis are the typical expressions of animism.

While exploring the 'variations of metamorphosis', Descola emphasizes how easy and obvious the 'plasticity' of forms can be:

> Often enough, it is perfectly 'ordinary' human or nonhuman persons, that is, ones with no mythological antecedents, that are credited with this capacity of metamorphosis, thereby testifying to the normality of the interchangeability of forms among all those who possess the same subjectivity. However, this plasticity is not total, and some modes of embodiment are less frequent than others. Conversion from animal to human and from human to animal is a constant feature in animist ontologies: the former process reveals interiority, while the latter is an attribute of the power with which certain particular individuals (shamans, sorcerers, specialists in ritual) are credited, namely the power to transcend at will the discontinuity of forms and adopt as their vehicle the body of some animal

> species with which they maintain special relations. The metamorphosis of a human into a plant or of a plant into a human is not so common and even less common is that of an animal into another animal species. [...] The conclusion that may be drawn from this list of possible and impossible metamorphoses is that the common fund of interiority stems from the set of characteristics observable in human beings, while the discontinuity of physicalities is modelled on the astonishing diversity of animal bodies.[17]

The reader of Ovid's *Metamorphoses* is bound to feel at home here. The possibility of metamorphosis haunts the poem. Human beings become stars, stones, trees, flowers and animals of all shapes and sizes. Nimbler than a shaman, Pythagoras has indeed lived as another person. Now, to say, as we did earlier, that this is a possible world implies that we, as well as the Roman readers in the first century CE, do not necessarily believe that these transformations could actually happen. Our own language supposes the standpoint of what Descola calls a 'naturalistic' ontology, namely the separation of the human experience of the world from the domain of nature. In our 'nature', a woman does not become a cow. We do not tackle here the broad question of the credibility of mythology (and of poetry) in ancient societies. But we must emphasize the contribution of the *Metamorphoses* to the contemporary reflection on a plurality of ontologies. Fresher than ever, Ovid's poem can be read as a thought experiment in animism. Whereas Lucretius denies both the possibility of new bodies coming to life, except through specific generation, and the existence of an immortal soul, Ovid takes up the challenge of these two postulates: human animation survives through infinite change.[18]

Reading animism

The *Metamorphoses* is a manifesto of animism. Whereas a structural analysis would scrutinize oppositions and reversals from one episode to another, an ontological reading brings to the fore the possibility of metamorphosis as if this were the essential meaning of the poem – what truly matters. Animism, however, assumes an idiosyncratic shape. It carries a reflective overtone. Ovid infuses his palpitating stories with an inquisitive meditation about the consequences of animation. Which nods to the readers. The potential for ecocriticism lies in this normative perspective.

In the past, metamorphosized bodies have been ostensibly human like ours; now, they have ceased to look like us. On the one hand, it is human beings who are in danger of being altered. *We* are concerned. On the other, the loss of anthropomorphic features is not a complete dehumanization. *One of us* is still, albeit invisibly, there. The relentless account of change acts as a revelation: we learn that differences have a history. Our fellow humans such as Io, Acteon, Niobe, Daphne have become others and, more precisely, imperceptibly hybrids. The text emphasizes imperceptibility over otherness. Some human subjectivity remains there, still. This storytelling, as we have said, generates a perceptual challenge: now we know that the universe is filled with shapes that are no longer human*like*, but in which a residue of human consciousness may go unobserved. Potential confusion, we should add, is contrived by the wicked deceptiveness of these hybrids, so beautifully emphasized in the poetic writing. If they were monstrous, like centaurs or other Empedoclean compounds, then it would be easy to spot their maimed humanity. But in Ovid's metamorphic world, Io is a perfectly bovine cow, Acteon is a handsome deer, Niobe is a solid rock, Daphne is a credible laurel tree. While preserving their human identity, a metamorphosis has surreptitiously concealed the human ego, which is *also* there. These 'also-humans' have become unrecognizable as such.

This inevitably affects the experience of this storytelling on the receiving side. To be sure, the perceptual challenge concerns, first, the characters involved in the stories, but it has also a disturbing effect upon those readers who immerse themselves in these stories and emerge on the other side. They – we – are humans. Metamorphosis after metamorphosis, something new is somebody lost. What is at stake is not merely Naso's virtuosity – let him write one more process of solidification of a woman into roots, trunk, bark, branches and leaves or the dilution of a boy into a spring, and so on and so forth – but the vulnerability of the human *forma*. We witness the vulnerability of *our own* form. Io's father or Acteon's dogs fail to detect the lingering presence of those young persons: but what about us? Perhaps, we too have something to learn from their predicament.

Storytelling is a manner of thinking. If I think that all things are in flux, then recounting how new bodies have come to be (and may keep on becoming) is the most adequate channel to convey this very idea. The medium does fit, mimetically, the message. Reader, do understand: once upon a time.... The narration of 'who was who' before being changed cannot leave us philosophically unscathed: not only does it present us with a liquid, animistic ontology, but by revealing the human origin of non-anthropomorphic beings, it insinuates a systemic doubt about our own sensorial faculties. Who might still be there, lurking in the new bovine figure now grazing before my eyes and trying to articulate words while mooing? One of us. This oblique provocation vis-à-vis the readers becomes explicit at the very end of the poem.

In Book 15, Pythagoras of Samos addresses a long, chastising speech to a general audience of ordinary people. They are all culpable of disregarding the fluidity of life. They murder and eat beings that are invisibly human. And now, with Pythagoras speaking, what started as a tale that mimics a fluid ontology – 'Let me tell you how things materialize over time!' – becomes a proper lesson: 'Let me explain how you must live your life!' Metempsychosis is the continuation of metamorphosis in our own world. Fourteen books of transformations bring about the urgency of giving up bloody sacrifice, meat-eating and any form of carelessness in our relations with seemingly non-human, yet also-human, beings. The misleading trompe-l'oeil, exposed in thousands of verses, still threatens Augustan Rome – every time people dine on a juicy stew, every time they slay a fine-looking bovid for a god, with the blessing of the Prince. The *Metamorphoses* acts didactically in its own way: it teaches the Romans alimentary scepticism.

Pythagoras' speech creates a truly special effect on our reception of the entire poem. It bridges the marvellous content of the epic and our own 'Roman' world. With Pythagoras, changeability overflows from the poetic storylines, and comes to concern us. We know that we are not arboreal nymphs, Arcadian kings, or Colchidian princesses, but we also know that, for the simple reason that we must die, we are entangled in a network of consanguinity. No exception. Metempsychosis is a credible threat for all – in and out of the text. We, readers, are left with a disquieting message. Eating meat might be devouring human flesh, therefore we can become – as unbeknownst to ourselves as Thyestes feeding on his dear children – cannibals. We must find an exit strategy: to convert to vegetarianism. And we must also rethink critically the murderous ferocity of that sacrosanct Roman ritual: animal sacrifice.

From Pyrrha and Deucalion up to Augustan Rome, it is this enfleshed manner of being in the world – humanity – that becomes soluble in the flux of things. We, reading beings, are confronted with the possibility of being changed, not merely as an ontological surprise, but as a tragic risk: anthropophagy. Now, all the readers of this poem, written at the beginning of the first century CE, are culturally situated in their own contemporaneity. This creates multiple hermeneutical options. One possibility is, for modern interpreters, to experiment with a hyper-contextualized anthropological approach. Giulia Sissa takes up this challenge by focussing on perceptual uncertainty, blood, meat-eating, animal sacrifice and the vegetarian antidote to the peril of cannibalism. To pre-empt the unbearable consequences of a liquid ontology, we must eat only non-metamorphosed plants. This is how the poem cleverly immunizes itself against the side-effects of metamorphosis.[19]

Anthropocentric, post-human and eco-critical in its own way, Ovid's poem, Sissa argues, is both animistic and Roman. By shedding a critical light on war and ritual immolations, this disingenuous poetry conveys deep qualms about Augustan society, as Sandra Fluhrer also shows in demonstrating the similar perspective that Ovid casts on ancient agricultural practices. A discourse on the banality of violence runs through the *Metamorphoses*. A very different hermeneutical option is to zoom in on the poem's 'textuality' and its macro-structure. This is the direction Marco Formisano takes in his essay. Between the coming into existence of countless bodies from the sludge left over by the flood at the very beginning, and Pythagoras' speech at the very end, we encounter the noir figure of Medea. If we resist the impulse to look for intertextual echoes, we can stare at her here and now. She is the magician, the human-and-divine superwoman uniquely able to master metamorphosis. Right in the middle of the poem, she stands out. She emerges, she intrudes, she brings back the amorphous mud of the origins. 'Muddy Medea' is a gush of materiality popping up to annihilate everything Roman and properly feminine: marriage, the family, motherhood. At the hands of an infanticide, this repulsive Earth completes Zatta's picture of a caring earthly maternity. It also resonates with Sandra Fluhrer's reading of the foundation of Thebes in Book 3. Cadmus kills a dragon and sows its teeth but Earth revolts, changing him into an even creepier snake. The poem, Formisano argues, makes place for an anti-mother – a dark, dirty, bloody Gaia. While pursuing seemingly diverging lines of interpretation, these contributions draw attention to the tragedy of metamorphosis.

Multispecies ethnography

If Descola's approach opens up one way of breaching the foundational nature/culture binary of modern anthropology, developments in the related field of ethnography open up another: multispecies ethnography has emerged as a new subfield of this discipline, intent on breaking down species barriers by charting the cross-species encounters of humans with other kinds of being. It offers a suggestive framework for exploring the more harmonious interactions between human and other species that make up another aspect of the *Metamorphoses*' ecological vision, and which the poem invites us to consider by incorporating a degree of ethnographical lore for each new species at the moment of its emergence. Multispecies ethnographers' interest in cross-species intersubjectivities is also grounded in a dissatisfaction with contemporary philosophy's solutions to the question of the subject of philosophy in the wake of its own reckoning with the problematic legacy of Descartes, whose brand of human exceptionalism was predicated not only on the isolation of humanity from 'nature', but on a particular view of humanity that excluded significant human populations.[20] More recent attempts to formulate the subject of philosophy in ways that seek to overcome the historic human/non-human divide, continue to perpetuate some of these forms of radical exclusion.[21] Derrida's essay on his naked encounter with his (female) cat, for example, has little to say about whether his shame before her has anything to do with her gender, and his fantasy about returning to the primal scene of Adam's first (shameless) encounter with the animals in Genesis is predicated on the exclusion of woman, who does not exist yet, along with the feeling of shame that her creation makes possible.[22]

It is in part for its more ecumenical understanding of humanity that recent environmental criticism has embraced multispecies ethnography as a preferable route into the question of cross-species intersubjectivities. Admittedly, many contemporary accounts of this genre of writing emphasize how it evolved *against* the traditions of human ethnography, its emphasis on relations across species a deliberate challenge to the anthropocentric premises of ethnography's prior life as the fieldwork of social and cultural anthropology.[23] But multispecies ethnography may also be seen as an organic outgrowth of anthropology's ethnographies of human cultures, which themselves evolved in the twentieth century from being modes of reducing non-Western cultures to objects of observation, and thereby reinforcing subject/object, culture/nature hierarchies, to being more reciprocal procedures, accountable to postcolonial qualms. The presumed authority of anthropology's Western observers, their capacity to offer objective representations of 'other' societies, gave way in the postwar period to a critique of the privileged, Western standpoint that such representations assume.[24] Out of this 'crisis of representation' new ethnographical procedures emerged: the self-reflexive 'field account', for example, deconstructs the participant/observer hierarchy by replacing the rhetoric of experienced objectivity with autobiographical accounts of the observer's experiences in the field,[25] or by staging dialogues with her informants.[26] As James Clifford highlights, the aim of dialogical accounts like these is not to represent a particular society or culture, but rather to locate cultural interpretations in reciprocal contexts and to render cultural realities as negotiated and intersubjective.[27] Multispecies ethnography, with its interest in how different species relate and adapt to other species with which they coexist, is a direct descendent of this relational development in human ethnography; indeed, as Donna Haraway would have it, the different species that coexist in their relational assemblages do not precede this encounter with each other but rather become the very species they are as a result of it. Ovid's poem concretizes this abstract idea. Both Emily Gowers and Francesca Martelli build on this insight in their contributions to this volume, by demonstrating the extent to which human histories and chronologies gain their meaning by intersecting the durations and cyclical temporalities of other life forms.

A further development in human ethnography to which Clifford himself contributes is a newfound emphasis on the fictional, even poetic character of ethnographical accounts, and a renewed attention to their textual status.[28] As he points out, the crisis of representation that ethnographers in the postwar period were grappling with came as much from a post-structuralist hesitation about the limits of representation as from postcolonialism's critique of the politics of attempting to represent the Other. One consequence of this hesitation was to recognize the degree of poesis and invention involved in the ethnographer's account: no longer viewed as a neutral redaction of a particular society's cultural script, the ethnography came to be seen as the product of a particular set of rhetorical effects, one that made use of many of the devices – narrative, metaphor, figuration – that characterize literary, even fictional, texts. If these effects are characteristic of human ethnographies, they are no less a feature of multispecies ones, which frequently bring artists into dialogue with scholars in order to formulate new ways of imagining cross-species interactions.[29] By enlisting Ovid in the task of explicating such interactions, this volume participates in this very endeavour; and by featuring John Shoptaw, a contemporary poet, among our contributors, this volume acknowledges the vital work that poets and artists do in reimagining (and helping others reimagine) those interactions, including other species' reactions to the human. Indeed, even before we consider the more complex devices of figuration associated with poetry and the visual arts, we find that poets can alert us to more basic questions of cross-species translation, as Shane Butler (2019a) highlights in his discussion of the pseudo-Ovidian 'Elegy on a Nightingale'. This poem, which catalogues birds (as well as other animals) along with the sounds that they make, attempts not only to transliterate bird sound into Latin for the benefit of its human (Latin-reading) readership, but also follows the invitation posed by Jakob von Uexküll to imagine the perceived *Umwelten* of other species by considering what the birds make of our voices. How might a nightingale hear this very poem?

The most important use that multispecies ethnography makes of metaphor and figuration is in its understanding of species itself, which exists as a shorthand figure for a more complex constellation of actants. The first answer that Donna Haraway gives to the guiding question of *When Species Meet*, 'Whom and what do I touch when I touch my dog?', is the figure of 'Jim's Dog' – a photograph of a burned-out redwood stump, covered in moss and lichen, in a shape that bears an uncanny resemblance to a dog.[30] The product of digital technologies, biological organisms and economic histories of human land use and leisure, the canine figure of 'Jim's Dog' helps Haraway make the point that the species designation 'dog' is only ever a shorthand for the corporeal interface of myriad actors and agents with their own narratives and histories. More than this, though, this figurative dog demonstrates how contingent such species categories as 'dog' are, constituted by humans to fulfil their own sense-making needs. Ovid makes a similar point in the numerous examples of transformation that make use of different species to give form to something else altogether. Hecuba's transformation into the dog-like rock formation would be the obvious example to adduce along the lines of 'Jim's Dog', and demonstrates how closely Ovid's appreciation of the formal (as opposed to 'essential') properties of species aligns with Haraway's understanding. His account of the different star constellations – *Ursa Major* and *Minor*, for example – would be another.

Science and wisdom traditions

If the ethnographical quotient of multispecies ethnography is indebted to postcolonial developments within anthropology, so too the species discourses that it understands have been subjected to various forms of postcolonial critique. The urge to taxonomize the world's contents may be seen as a symptom of Western science's colonial project of exploration and conquest – a means of imposing a singular (western) classification system on the world's variety of living organisms without regard for local terms, narratives and systems. Against this impulse, the natural histories of ancient cultures, including those of Greece and Rome, present a refreshing antidote. Geographically centred though they may be in the world of the ancient Mediterranean, they combine close observation of flora and fauna with mythological lore about the human and/or divine origins of these various other life forms, and see no contradiction between the two modes of explanation. It is precisely for this reason that Bruno Latour, in his landmark study of the epistemic polysemy of modern science *We Have Never Been Modern*, suggested that contemporary science was in fact far closer to the intellectual range of science of the so-called 'premodern' period. As Latour points out, and as Miranda Griffin's paper in this volume deftly illustrates, ancient and medieval thought does not separate off science from mythic lore and other kinds of knowledge and narrative, but makes each domain of knowledge interpretable – even comprehensible – in light of the other. Griffin demonstrates that for the Medieval French authors and illustrators of the *Ovide moralisé* and other contemporaneous bestiaries, 'science' encompasses all that a text like Ovid's *Metamorphoses* has to teach – about cosmology, but also about how best to lead a pious, Christian life. As we look to figures like Latour, who straddles multiple forms of science (social and otherwise), for guidance about how to live with and respond to the ecological problems of our own times, so medieval authors looked to Ovid as a philosophical sage to help guide them through their own difficulties and inquiries in the name of science.

The mode of epistemic flexibility that Latour advocates, and which Ovid's poem, and the tradition it spawns, showcases, is not limited to domains and categories of knowledge but extends to geographical regions and cultures as well. Roman myth, like Roman religion, is famously inclusive, incorporating the legends and deities of the lands that the

Roman empire incorporates into its own story systems. Ovid's poem bears witness to this process of cultural hybridisation, and spawns a literary tradition that is highly sensitive to its ecumenical approach to cultures as well as to the notion of 'science.' Julia Lupton's paper in this volume reveals what inspiration Shakespeare drew from the geographically and culturally expansive tradition of 'wisdom literature' on human/other-than-human relations that Ovid incorporates into the *Metamorphoses*, and how, in *A Midsummer Night's Dream*, Shakespeare himself expands on it, incorporating references to Indian and Jewish lore into his homage to Ovid's poem.

The potential for this ethical, moralizing reception of the poem can likewise be seen in Claudia Zatta's discussion of the cosmogonic deployment of the ecosphere promoted by Arne Naess, who, however, diverges from Latour in understanding a starker separation of ecosophy from science.[31] Ovid's poem aligns with the principles of Naess' deep ecology, in showing us how the primeval, peaceful distribution of beings in multiple natural spaces is connected at times to a fundamental analogy between humans and plants, and, as a consequence, to the harmonious relations that analogy calls for. Like Earth herself, Dryope, an affectionate mother transformed into the lotus tree, never ceases to call for compassionate respect, *pietas*. So too, in the more recent history of ecosophy, Arne Naess has inspired a movement of people seeking to ground their practice of living in principles drawn from observing the relational needs of other life forms.[32] In her paper, Claudia Zatta demonstrates the commonality that Ovid's poem shares with this philosophy, as the analogies drawn between earth, plant and human, for example, encourage harmonious relations between them.

Agriculture and food studies

Ovid's poem is unusual within the field of Latin literature for its interest in exploring the interpenetration of human and other beings to the degree (and in the way) that it does.[33] More often the relationship between the human and the material world that Roman authors are at pains to stress is one of human exploitation and control. Farming manuals, which comprise a prominent literary genre in Roman culture and, as Diana Spencer's and Sandra Fluhrer's papers show, an important cultural background to the *Metamorphoses*, present an obvious example of this attitude, insofar as they draw a direct equation between the elite landowner's status of dominance as general of an army, master of a household of slaves and farmer of the land he owns. The extent to which the Roman agronomists devote their treatises to the management (or control) of slaves – slaves who will manage (or control) the land and farm animals on their behalf – makes the analogy with their control of 'nature' inevitable. For it is not just that slaves are reduced to the same object position as domestic animals and fields within the social hierarchy of the rural villa, but also that the agronomists' preoccupation with slaves lays bare the servile status that they assign to all the natural resources in their hands. As ever, when humans set about constructing nature/culture (or, as here, nature/agriculture) oppositions, these very distinctions are propped up on the invisible human populations that get swept under the mat of 'nature'.

If slavery is one aspect of Roman society that inflects the ancient agronomists' view of natural resources, Roman militarism is another. The emergence of the farming manual as a genre is commonly linked to changes in the agriculture industry in Roman Italy and to broader socio-economic changes taking place across the Roman world of the late Republic: as Roman militarism became professionalized, and the urban centres in Italy grew, the small-scale farms of Roman citizen soldiers became incorporated into larger latifundia, in which agricultural production was managed on a more industrial scale, better suited to serving the needs of the growing urban populations (or so the story goes).[34] The connection that the manuals make between farming and militarism is, on the one hand, a nostalgic one, a throwback to the imagined (and imaginary) days of Cincinnatus. Yet, as Spencer highlights as part of her survey of some of the chief social and cultural contexts for the *Metamorphoses*, the coincidence of the growing industrialization of food production in Roman Italy and the professionalization of the Roman army suggests parallels with the rise of modern agribusiness, the innovations of which are commonly linked to technological advances made in the weapons industry during the First World War.[35] Spencer demonstrates how important an influence the innovative techniques used to increase agricultural production in Rome are on Ovid's interest in hybrid beings.

Positioning Ovid's *Metamorphoses* against the history of this farming literature, as both Diana Spencer and Sandra Fluhrer do, is instructive because of how infrequently the theme of agriculture appears in the poem. The story of Proserpina, which provides an aetiology for the plough in Ovid's *Fasti*, is evacuated of this function in the *Metamorphoses*, and although seasons and harvests remain an important object of concern in this telling of the myth, they are presented as the spontaneous gift of Ceres and dissociated from human agriculture throughout. One of the few stories in the *Metamorphoses* that does display traces of agricultural activity is the story of Cadmus' foundation of Thebes in *Met.* 3. As Fluhrer's chapter in this volume demonstrates, the sowing of dragon's teeth, and the violence that ensues from that act of cultivation, presents warfare and agriculture as a seemingly inevitable entanglement from the outset, in ways that evoke the military inflections of the narrative of Roman agribusiness set out above. Yet in this story, as Fluhrer also demonstrates, it is the land that fights back, resisting the process of cultivation and acting upon the human agents who attempt to impose their will upon it. This effect is reinforced by the somatic effects that the conclusion of the story (Cadmus' final transformation into a snake) has on the reader, a visceral reaction that at times reflects the bodily experience of Cadmus himself, as we feel the horror of his changing form on the surface of our own skin, in spite of our own reasoned knowledge that this is just a fiction. In this story, as in that of Medea, matter – the material earth, the material body – resists, recoils, reacts.

Hyperobjects and apocalypse

The resistance that we see the earth mounting in the Cadmus story is precisely a refusal to be subjected and dominated (as per the Roman farming manuals), and a measure of her status as an actor in her own right, a constitutive element in the assemblage of life forms and other actants that Ovid presents throughout the *Metamorphoses* as an interlocking system. Bruno Latour's preference for thinking in terms of assemblages of disparate actors offers a guide to doing away with subject/object structures and hierarchies, and, for many contributors in this volume,[36] presents a favoured rubric for understanding Ovid's multispecies (and/or multi-actor) encounters. But these assemblages are not always evenly distributed, and there are times when the network of actors gives way to an entity that dwarfs the other elements. Contemporary ecocriticism has coined various terms to describe such phenomena: co-opting the object-oriented philosophies of Graham Harman and others, which posit the autonomous existence (and metaphysical reality) of objects, independent of human perception, and their 'withdrawal' or capacity to exist in excess of their relations with humans and/or other objects, Timothy Morton coined the term 'hyperobject' to describe a particular kind of object that insists on its own objective and objectifying force. The hyperobject may not be said to exist as a result of its validation by human consciousness; yet one of its most distinctive and paradoxical properties, is precisely the effect that it has on humans, as a result of its looming, yet frequently invisible, proximity (as indicated by the prefix 'hyper'), which 'reduces' the human to the status of another object in the wake of its irresistible force. Hyperobjects are things that are so massively distributed in time and space that we cannot perceive them in themselves (or in their entirety), but only know them through their material effects. Entities that stretch the limit of being described as such – like the biosphere, or global warming.

Early on in the *Metamorphoses*, we encounter two episodes that serve as obvious ancient parables for the modern phenomenon of climate change, and for which the hyperobject provides a suggestive tool of analysis because of the warping effect that these climate parables have on the human and other objects that run up against them. After the account of creation at the start of *Metamorphoses* 1, the reader is soon confronted with the spectacle of this physical order giving way repeatedly to disorder again, with the Flood myth of Book 1 and the story of Phaethon in Book 2. In the case of the Flood myth, the displacement of creatures to elements and geographical locations in which they clearly do not belong renders them helpless and, along the way, discloses their own object status when placed within the engulfing climate event that Ovid describes. Object-oriented philosophers see the world in terms of interobjective, rather than intersubjective, relations, and argue that it is the function of the hyperobject to flip our anthropocentric biases and to disclose the very premise of the subject as an effect of relations between objects.[37] These relations differ from intersubjective ones by virtue of the impassive view that they take of the actors involved, the systemic nature of the interactions between them and the emphasis that they place on cause and effect over intention and meaning. Ovid's account of the Flood enacts this subversion of intersubjective relations: it directly reverses the terms of the creation narrative, wherein the assignment of particular creatures to particular locations was determined by the intentional activity of a rational demiurge, a quite precise figure for the intending subject, whose actions furthermore separate out the creatures, allocating to each of them a distinct category of being by assigning them to their own particular domain. The disorder of the flood, by contrast, while ordered by Jupiter, is produced by a change in climate, a phenomenon that was hitherto unseen in the poem, but which we now perceive in the chain of displacements that it effects, as entities that were separated off are now shown to interact with one another in an interrelated system.[38]

In the Phaethon myth that arises at the start of *Met.* 2, the world is destroyed not by flood but by fire in another parable for climate change that John Shoptaw draws out in his poetic rendering of the episode, inspired by recent wildfires in California. The allegorical status of this myth has an ancient pedigree: it was used in antiquity to illustrate the Stoic idea of ekpyrosis, according to which the earth is periodically destroyed by fire before being recreated anew. Ovid's version of the myth is strongly focalized through Phaethon's eyes, and complements the earlier myth of cosmic destruction by flood by emphasising the human experience of the hyperobject. Phaethon's terror at the global conflagration he creates reveals the abyssal status of the hyperobject relative to the human, and the attendant feelings that it generates in the human actors who run up against it, as they experience its objectifying force. Phaethon's pettish concerns about paternity, which catalyse the disaster, allow him to believe that he has the birthright to control an object (the chariot of the sun) of unimaginable force. The terror that he experiences as a result of his inability to curb the destruction that his loss of control wreaks on the earth, speaks not just to the sublime awe that the majesty of the natural world has ever inspired in the human, but to the peculiar feeling of helpless horror that the manmade hyperobject inspires in its human creators. Shoptaw's version of this episode offers a different twist on the affective valences that attend the event of global warming. His Phaethon (named 'Ray' and modelled on the climate-change denying Donald Trump), is distinguished not by undue awe at the conflagration he has created but by a callous disregard for the havoc his actions cause. The objectifying power of the hyperobject is, in this instance, manifested in the object-like, impassive reaction it inspires in the human 'Ray', who both brings it into being and is subsequently destroyed by it.

Conclusion

Ovid's epic poem displays a porous, malleable cosmos, in which humans may disrupt the earth's balance, even as their own human form is itself disrupted by their transformation into other living beings. This is the ontological strategy of the poem. But the poet has thought of us, the human readers. We sit outside the possible world of the poem, but as human beings, we can relate to the potential victims of metamorphosis, in the text. More poignantly, as mortals whose souls, if Pythagoras is right, die and move on from body to body as a matter of course, we are liable to reincarnation – and to carnage. This is why the poem brings to the fore not merely the welcome, oecumenic, equalizing inclusiveness of an interconnected ecosystem, but the dark side of ontological fluidity, more precisely its virtual tragedy. The papers in this volume aim to extrapolate both sides of Ovid's ecological vision: the nodes of differentiated life that sustain the biosphere, as well as the modes of incursion that threaten to make its distinctions collapse.

Notes

1. This volume emerges from an international conference 'Metamorphosis and the Environmental Imagination', held at UCLA, under the auspices of CMRS-CEGS (Center for Early Global Studies), 11–12 October 2019. A kindred interest in applying contemporary environmental criticism to ancient texts can be found in Schliephake 2016; and Schliephake 2020. See also the panel dedicated to 'Ovid and the Natural World', at the Society for Classical Studies meeting of January 2022, under the direction of Laurel Faulkerson and Carol Newlands.
2. See Zatta 2016; and p. 167 for Ovid's engagement with the Presocratics.
3. In its interest in the interpenetration of 'nature' and 'culture' (or, as Haraway 2003 puts it, 'naturecultures'), Ovid's poem diverges from that tradition of ecocriticism which laments the alienation of the human from the natural world, and thereby understands 'nature' as a discrete metaphysical category, a tradition that arguably begins with Virgil's *Eclogues* (with Armstrong 2019) and is best represented today in e.g. McKibben 1989. Ovid's poem belongs in a more posthumanist tradition of ecological writing, which focuses on the entanglement of the human with other life forms, a tradition treated in modern Classical scholarship by e.g. Payne 2010; and Brill, Bianchi and Holmes 2019. See Heise 2016: 8–10 for a useful survey of these divergent ecocritical traditions.
4. See Chesi and Spiegel 2020 for a broad application of current trends in critical posthumanism to the field of Classics.
5. We follow Morton 2007: 8–14 in preferring the term (and project) of 'ecocritique' to ecocriticism, for its calling into question of a reified idea of nature and a concept of the environment as a locatable, exteriorized entity. Ecocritique thinks critically about some of those aspects of environmental discourse that have, unwittingly, reinscribed the object status of the systems it seeks to defend. The genre of 'nature writing' is, according to this account, one of the worst culprits.
6. See Spencer's essay in this volume, however, for an important discussion of the hybrid discourses on display in the texts of Roman authors (Cicero, Catullus, Virgil) of the generation immediately preceding Ovid, which provide important context for the discursive fusions of the *Metamorphoses*. For discussion of how the poem combines scientific with other poetic discourses, see Myers 1994; and Martelli 2020: 38–42 and 53–5.
7. Ov. *Met.* 15.178.
8. Ov. *Met.* 1.1–2.
9. This is an allusion to a significant trend of literary theory, loosely called 'reader response'. On Umberto Eco's theory of interpretive 'cooperation', see Pisanty 2015.
10. Ryan 1992: 537: in fairy tales, 'the supernatural is experienced as natural by the characters and the possibility of the impossible is taken for granted'.
11. Ov. *Met.* 15.183–5 : 'So time both flees and follows and is ever new. For that which once existed is no more, and that which was not has come to be; and so the whole round of motion is renewed. (*Tempora sic fugiunt pariter pariterque sequunturet nova sunt semper; nam quod fuit ante, relictum est, fitque, quod haud fuerat, momentaque cuncta novantur.*)'
12. Hardie 2002a.
13. In his own *Metamorphoses* (2.1–2), a novel intertextually connected to Ovid's *Metamorphoses*, Apuleius has Lucius experience this kind of doubt in Thessaly.
14. Lévi-Strauss 1962, 1964, 1966, 1968, 1971. The most serious attempt to interpret Graeco-Roman mythology, including Ovidian texts, along these lines is Detienne 1972.

15. Frontisi-Ducroux 2017.
16. Descola 2005 (English trans. 2013). In the same perspective, see also Pitru 2017.
17. Descola 2013, 73.
18. Lucr. *DRN* 5. 821–924. See Campbell 2003.
19. Sissa 2019, and pp. 35–54 in this volume.
20. Cf. Plumwood 1993, 107 for discussion of how Descartes' project participates in the colonial project of casting colonized peoples as primitive, and thereby closer to nature, in contrast to the peoples of the 'civilizing' West, thereby perpetuating a model of humanity seen in Aristotle and Plato, which denied the enslaved the human *ratio* attributed to citizen males; and on how it continues to attribute inferior degrees of consciousness, and therefore humanity, to women, whose reproductive and servile functions are also assumed to bring them closer to nature.
21. See also the critique by Haraway (2008: 27–30) of the misogynistic premises of the account by Deleuze and Guattari 1987, 232–309 of the post-Oedipal subject as the end product of a process of 'becoming animal' and even 'becoming female'.
22. Cf. Derrida 2008: 50–1 for the culmination of this fantasy. See also Shane Butler's paper in this volume, which furthers Derrida's inquiry into what the cat sees when it looks at a human, by asking what birds hear when they hear humans speak or sing.
23. Cf. Kirksey and Helmreich 2010: 550.
24. Clifford 1986: 10 offers a helpful summary of the emergence of this critique in the postwar period, which arose not only in the wake of Said's critique of colonial strategies of representation in the west but also in response to Ong's critique of the West's privileging of sight over the other senses in its scrutiny of the Other.
25. Clifford 1986: 13–14 on the self-reflexive fieldwork accounts of e.g. Beaujour 1980 and Lejeune 1975.
26. Clifford 1986: 14–15 on the narrated dialogues of Lacoste-Dujardin 1977, Crapanzano 1980, Dwyer 1982 and others.
27. Clifford 1986: 15.
28. Clifford 1986: 3–8 on the use that ethnographies make of literary tools to describe behaviours that may themselves be called forms of cultural poiesis.
29. The work of the Multispecies Salon is exemplary in this regard: https://www.multispecies-salon.org.
30. It can be no coincidence that the Jim of 'Jim's Dog' is none other than James Clifford, author of much of the ethnographic work cited above, and Haraway's colleague at Santa Cruz. Clifford's influence on Haraway's thought is apparent from the numerous citations of his work throughout Haraway 2016.
31. See Zatta (p. 163 below) on the distinction that Naess himself draws between the principles of deep ecology and those of hard science.
32. Note, however, the important critique of this movement by Guha and Alier 1997, who take issue with its privileging of wilderness, as a discrete metaphysical category, over environmental justice for impoverished human populations in the global South.
33. Although both Lucretius' *DRN* and Virgil's *Georgics* explore different forms of interpenetration between human and other life forms, Ovid's poem foregrounds this process by dwelling on the literal, physical transformation of humans into other animal, vegetable and mineral forms.
34. Cf. Fluhrer (p. 202 in this volume) for bibliography and discussion.

35. Cf. Carruth 2013: 12–13.

36. E.g. Griffin (pp. 126–44); Formisano (pp. 58–9); Fluhrer (p. 216); and Martelli (p. 99).

37. Morton 2013: 84.

38. Of hyperobjects, Morton 2013: 85 writes: 'Since we only see their shadow, we easily see the surface on which their shadow falls as part of a system that they corral into being'. In the Flood myth of *Met.* 1, that system (here, the food chain, or even the biosphere) is only revealed by a disruption to the conditions of its existence – that is, by a hyperobject that effects its dissolution.

WHOA!

Fired by dreams of the sky, Phaethon, son of the Sun,
leaves Africa, crosses star-scorched India
and eagerly draws near his father's rising.

Ovid, *Metamorphoses*

One

A young man with gold hair
in a coal-black robe and slippers
was off to confront the Sun. But
as he paced his hotel corridors, Ray
could feel his step losing its jaunt.
At this rate, he'd make it to nowhere
in nothing flat. Just then, he noticed
his old wall map thumbtacked
over some double doors. How'd his
Boys' Life get out here? He looked it over,
the big fat world, cut open and stretched
flat, like an elephant hide. The properties
were green, purple and red, and the waters
whitish blue. The Mercator-projected
bottom was jagged with ice. The whole place
was crawling with letters, like ants on a cake.
The best part was the creatures: a long-armed squid
suctioning a whale, sea lions sunning
themselves. No. Seals. He bunched
his brows. No. Hills. And game
big as states, as nations: hippos
in Africa's watering holes, a beauty
of a cheetah, its sights trained. Untacking
and folding the map, Ray tucked it
into his pocket and slid open the doors.

He found himself in a dim cavern
facing a motionless escalator. Its stairs
were gold-plated, its balustrades brushed aluminum
and its handrails hand-rubbed ostrich skin.

He could see no end of stairs. It must be
one unbroken flight to the Sun's penthouse.
But where was the down escalator? Beats him.
He planted his slippers and clasped the rails.
With a little lurch the escalator began
its quickening glide. Now this was first-class.
No huffing required. Simply the assumption
of an attitude. Up ahead, the horizon
gleamed ivory.

Two

The stairway teeth
clamped shut and Ray was thrown
into a hard glare. *Sunblock and goggles –*
they'll let you look into things – on your right,
came a voice positively sunny.
Ray slathered and goggled himself,
and beheld by a high stone robed
in emerald moss a trim figure
in a sea-snail-purple sweatsuit
and a broad-brimmed straw hat
puttering about a humid sundial
of wide-waking annuals and perennials –
snowdrops, crocuses, daisies and daylilies,
rice in flower and maize in silk,
woozy jasmine and heady grapevines,
a temperamental winter daphne, and tall
in the motley midst the polar gnomon,
an Indian turnsole or heliotrope,
sweeping its shaft of shadow from one
to the next minute marigold dot
as the ambling photosynthesizer passed
the fragrant hours of the zodiac. Batting
away hummingbirds and honeybees, Ray
waded through the flower bed toward Sol.
Am I your heir? Can I have what I want?
Sure, kid, name it. Ray glanced
toward the honey-scented manure.
A joyride? Sol sighed. *It's your funeral.*
Taking his time, he led Ray
out to the dayport.

What a letdown!
Ray'd expected a futuristic helium-
powered sports coupé, or at least
a souped-up solar Mustang. Not this
gold-rimmed, pyrite-axled
backward half-bucket with footpads!
Dawn, meanwhile, with her practised blush
arched her horizon, and her pomegranate interiors
yawned wide. The stars dispersed.
Venus, her shell-blue and egg-white gauzes
aswirl, gave Dawn a last long look
above her handmirror's crescent rim.
Sol was watching moonshine evanesce
from his glowing barndoors, which cracked
in a spewing welding arc and then
opened wide to reveal a dozen prompt
and close-cropped Hours, coaxing a team
of four mammoth pegasi yoked
neck and neck and neck and neck,
stoked with thick ambrosial plasma
and snorting flame. As the nymphs hitched
the tack to the chariot pole, the Sun,
shaking his head, untied his flaring
sunhat and knotted it under Ray's chin.
Lay off the whip and draw rein
with a firm hand. They'll fly by themselves
with no help from you. But they'll answer
to Clover, Crocus, Peony, Sage.
Steer clear of the straightaway. Bank
the rig into a circumbendibus. But don't,
whatever you do . . . But Ray's attention
had drifted. The rampant quadrupeds hooved
trails of sparks down the dayport door,
their heavy fog blankets slid off,
and they shook out their shoulder-wings.
A pliant breeze sprang up from the East.
The Hours passed with worried glances.
Ray hopped onto the nimble chariot
(his small feet lost in Sol's footrests),
He jingled, and then had to lunge for
and cling to the lines used by Sol
to hold in check the intent four-in-hand.

Three

The team shot up toward a rosy ice-cloud
and melted through it. Then everywhere was blue.
What once was a tailwind now blasted Ray's face.
He peeked down behind him from the open chariot.
A purple speck was waving. Possibly.
What was he doing up there? His stomach
lurched. *Whoa! Back to the stables . . .*
Peggies? Horsebirds? He'd forgotten their names.
As when, out for a spin on a winter blacktop
in his dad's SUV, a full-grown boy
sails onto a lake of black ice, skids
away from a skid that keeps on skidding,
the horsepowered four-by-four starts driving
itself; just so, sensing they were charioteered
by a lightweight, the horsebirds relieved themselves
and bolted above the worn ecliptic
toward the upper atmosphere. The reins slipped
from Ray's numb hands. A cluster
of climate satellites constellated themselves
against the stampeding chariot – as a bull,
a polar she-bear, a scorpion flexing
its venomous pen. One by one,
their sensors were going dark. Then
as a sidewinding snake the satellites
spooked the horsebirds. The chariot plunged
earthward, underneath an awestruck moon
and burned a thunderstorm to a crisp.

Four

Sniffing the pastured earth below them,
the horsebirds pulled out of their dive and settled
into a scathing orbit. Only Ray
remained skittish. He inhaled a stale
heat as from an unvented greenhouse, where the air
had lost its breath. It got dark out as pine pitch.
Ray felt the white-hot footrests
through his slippers. Ashflow smeared his goggles
and shooting embers dinged the chariot.
Smothered in smoke, he had no idea

where he was or might be heading.
Then a sharp breeze cut the dark,
and Ray looked down on someplace shaped
like California, gored by flames. Coulters
and ponderosas, yellowed and browned, engraved
with trilobite grooves by pine-bark beetles
wintering northward, had turned from trees
into tinder. Hope in the Santa Ynez
mountains held out like a manzanita gripping
the porous ground till a Santa-Ana-
driven wildfire baked the soil, leaving –
after a rain heavy as lead – water beads
and clay and ash and boulders and mud
to smash down the watershed clear to the beach,
upon which Hope too gave way.
Ray's head was spinning. Through the smoke,
the horsebirds smelled saltwater and sailed
that way, out past where petroleum platforms
swarmed the shoreline like mosquitoes.

The Pacific swallowed what she could take
of the bad air. Then she downed some more.
She was warm because she was carrying La Niña.
But she also felt feverish. She belched
acid up into the heat. Ray unfurled
Boys' Life from his robe pocket
and leaned with it over the chariot rim.
He spotted Alaska, which seemed to have shrunk.
Like it'd been left too long in the dryer.
A heat wave rose from the Alaskan Gulf,
turbulent with the psyches of dead young cod,
and broke over the drilling state, toppling
records from Juneau to Barrow. Ray saw
nothing for a while but rough blue glints.
Then he watched a king tide empty
Waikiki Beach of vacationers who
surged back as the tide receded.
He counted on his fingers. Recounted. Funny.
There weren't as many Solomon Islands
as Boys Life swore there should be.
And when did Fiji become a desert?
Ray picked out the Marshall island
with the concrete plutonium dome – built

for the ages – lapped by a salt swell.
He smoothed his robe. He was getting the hang
of this sun-buggy. If only he could reach the reins!
Off the coast of Australia, where coal bulkers
rocked in port, Ray mistook
the bleached bones of the Barrier Reef
for the beached corpse of Moby Dick.
Where to next? The horsebirds were oaring
toward the globe's bright protruding backbone.

Five

Ray looked down on the Himalayas.
Everest, the planet's summit, looked
like it was climbing up to meet him.
But its exposed glacier was melting,
like waxed wings from Icarus's shoulders.
Another Himalayan peak, one *Boys' Life*
called Gangotri, had icewater trickling
down her flank into the floodplain of the Ganges,
where the goddess Ganga was laid bare
from the waist up in the middle of her sinking,
human-soiled river. She'd never imagined
her years were numbered. She scowled at Ray,
who averted his gaze. Nature bored him.
Where were the cities? *Boys' Life*
suggested Mumbai, New Delhi,
but Ray could see only chemical clouds.
Way up to his right, he spotted the fume
over Beijing. He squinted and his close-up goggles
pierced the smog. Their mouths cupped,
city-dwellers felt their ways, walking
by lamplight or riding by bike light.
Mute as statues, the elders bent
over their blinking youngsters, as though
to appease the local coal divinity.
The horsebirds swung the chariot north.
The Siberian Tundra braced itself
for spring. Ray picked up a desecrated
music swelling from a vast pipe-organ
carrying liquified natural gas
from unearthed plant and animal plankton.
The pipes branched from Russia into China,

Turkey, Syria, Ukraine and Europe,
where from a myriad of water heaters,
teakettles and stovetop espresso pots
issued a Calliope riot of clanks,
hisses, shrieks, and coughing fits.
Ray looked on the bright side.
He'd weathered the worst of it. The first world
lay ahead, snuggled in winter.
The Caucasian glaciers looked thin,
but winter snows would plump them up
like mashed potatoes. Everywhere life
was sleeping in or slogging on.

Six

I know, Heraclitus, everything streams –
even paralyzed waves of ice –
but how many times do you print your feet
in the same Meander silt at Miletus
till you know you've reached the dry mouth
of something? Ray sought his reflection
in the blue Mer de Glace and found it
stubbled with gravel and rubble.
The glacial sea was shriveling up
under Mont Blanc's blank gaze.
Ray's own wandered to the Dolomites
where neighing carriages hauled skiers
up streets of anthropogenic snow;
to the Rhône glacier, blanketed in white
reflective blankets by Swiss teams
reduced to hospice nurses, its darling
Lake Geneva's temperature rising;
to the Rhine glacier as it flushed its river
sparkling with microplastics; to the cracked
mud bed of the glacial Danube,
its fluctuations dammed, its timeworn sturgeon
floundering. Ray's goggles glazed
Boys Life into an awful nameslide—
Pasterze Hornkees Hintereis Ferner
Engabreen Storbreen Aletsch Griesgletcher
Ragujiekna Storglaciaeren Saint Sorlin
Sarennes Maladeta molten Aetna
stepping into the Mediterranean.

Seven

What I'd give for a swig from the Hippocrene
hiding somewhere up on Mount Helicon
where Pegasus sank his hoof and the water
bucked, or even a swallow of Arrowhead
from the San Bernardino Mountains bottled
in clear recyclable petroleum plastic.
The *Metamorphoses* looks down on its clueless charioteer
while its mountains and rivers unroll in perpetual
hexameters serene as a Chinese handscroll.
I'm not up to it. I have no command over
my recalcitrant materials, so far beyond my
power to remedy or return to their courses
that despite my getting a feel for my rendition
I know I'm also a petrified rider
whose classic vehicle is running out of fluids.
It's all I can do to follow the circumnavigation
to its downfall in my stamping four-horse verses
on a long day when it's always noon.

Eight

A woman in a fashionable winter coat
paused in the iconic Piazza San Marco.
Her face was hidden, but her body language
indicated to Ray that she was waiting, hip
deep in the Adriatic, for a table to be vacated
by the seagulls, who'd displaced the Piazza's pigeons—
unless she was posing in an art installation
to call attention to the sinking attraction.
Ahead, a hot-spell, slopping and exhausted,
sank onto the heat island of Paris,
awakening the brimming Seine which evacuated
quarters and museums, though Magritte imitations
popped up in the glimmering floodwaters in the form
of nouveau lampposts. Mindful of her glory,
Athens had filled her empty streams
with concrete and asphalt smooth as alabaster,
but floods flashed and thundered through her recent
developments. The Sun's son seethed.
What's with the flooding? Today was his

to letter gold, to burn something into nothing.
He brightened at the sight of the Po's drought plain,
at the Tiber slowing to a stately walk
past speechless Roman fountains,
at the molten Tagus stopping altogether,
its bare floor baked gold,
its hillside forests glittering with wildfire.
Europe teemed with exotic plant life,
a host of charcoal asphodels fattened
on aged and dry-pressed swamp nuggets,
and blossomed into weirdly colored plumes,
tinting the sun's rays and altering
every breath, a seasonless reminder
of the fruits of industry. Belchatów and Kozienice
bloomed carbon dioxide, Thyssenkrupp
particulate matter; cadmium, mercury
and lead blossomed from U.S. Steel
in Slovakia, from Arcelor Mittal in Spain
and from Gaia Littoral in Portugal; Seine
Aval bloomed waste-water nitrogen;
Scunthorpe Integrated Iron & Steelworks
was fragrant with polycyclic aromatic hydrocarbons.
Their incense rose to high heaven.
Then Ray caught a whiff of shame.
Europe was turning away from its smoke-stalks
and its tailpipe scents. They were harnessing energetic
winds and even the sun. Whose chariot was
it anyway? Those eyeless sunshades,
those propellers propelling nothing nowhere
annoyed Ray. In fact, he was fed up
with the entire continent. He wished like hell
he could pull out and head – horsebirds willing –
back to America. Instead, they steered
the chariot south toward Africa,
which was fine by him. It must be sizzling.

Nine

The desert indeed sizzled tan
and gold charred brown by the wind.
Then Ray saw something new
under the chariot that Boys Life missed:

corn and peanuts, cabbage and cauliflower,
oranges and grapes. Libya had tapped
into an ice-age aquifer from before the Sahara
and channelled it – through a jointed web
of well fields, pipes and reservoirs –
into a man-made river network guaranteed
to flow for thousands of years. (Some
said forty.) Ray looked east.
Egypt was all Sahara, apart from the Nile,
whose seven mouths gurgled seawater.
She shrank back toward her lake plateau
where things had always been cool and blue.
Till now. Blackened by drought, Ethiopia
was siphoning off her headwaters. Egypt
flexed his arms. Meanwhile, the Sahara
was swelling south into the Sahel, where rainstorms
metamorphosed into sandstorms, millet fields
into goatherd clearings and ana trees
into firewood. Like Midas, the groping desert
turned all it touched into sand. Then Ray
noticed a rustling green borderline,
a floodwall running from the Red Sea
to the Atlantic. People was nursing trees
and sprouting buried roots to check
the Sahara's migration into the drylands.
The drought, though, would claim what it could.
Cape Town was a minute or two
from zero hour. The thousand-year icecaps
of Kilimanjaro would run down in forty.
Then what? Would the forests with their monkeys
outclimb the heat till they got pinched
off the mountain tip? Who knows?

Ray was not one to dwell on consequences.
He searched *Boys' Life* for something to do
and he came across the Serengeti with its thumb-sized
wildebeests, zebras and lurking cheetahs.
He wished he'd packed his big-bore long shot.
Don't even think about it! the horsebirds bristled.
But down below he spotted only
gnus gnawing on savannah stubble.
And everywhere he looked he saw broad
and sluggish human streams deserting

Mali, Niger, Nigeria, Chad.
(Lake Chad near dry.)
Bony herders with their fly-stormed herds
of cattle and goats were flooding farms
and conservation islands; tribes fording
the Red Sea strait into Yemen were changed
in the crossing from emigrants into climate refugees.
Something's wrong. Animal migration
was seasonal; human resettlement permanent.
What then was this migratory flow of people,
animals, plants, sands and weathers
but an irreversible, unthinkable, dead-end migration?
Malarial mosquitoes rode the thermals
up into the dammed Ethiopian highlands.
Ray felt his skin creep, as
Time flew toward the circulating saltwater.

Ten

Soon as the chariot reached the Atlantic,
it was met by a wobbly wind. The horsebirds
cocked their heads and the car slid
sideways. Ray grabbed hold of the rim.
Something in the global conveyor belt
of the oceanic circulatory system had slackened
like the flapping reins of the solar chariot.
The cooling gulf-water migrating north
ought to have flipped and dived like escalator
stairs swimming in darkness back down
to their lower landing or like a bowhead whale
underneath her cracking polar pack-ice,
looking to dine on a mess of krill
with her consort in the unlit infernal deep.
But instead, by a freshly groaning Greenland,
the lightly salted and luke-cool stream
seemed afloat in second thoughts.
Northern sea-life also wavered.
Arctic terns, recirculating from Antarctica,
missed the sand eels that'd fled
warm weather; bottlenose dolphins
arced and bull sharks switched
and both washed up off Newport News

and New York; turtles stranded
on Cape Cod learned they couldn't
turn back into sea turtles; codfish
in the Gulf of Maine turned belly up;
and lobsters scuttled along the sea bottom
north toward Labrador. Ray didn't notice;
he'd just spotted Midtown. High
above the vacillating Gulf Stream
the jet stream sagged and snapped
New York and New England
into winter blizzards. The fossil lordlings,
their attorneys and traders brushed it off
as they'd done with Neptune's trident
of overheated tropical turbulences – Harvey,
Irma and Maria. They went out
sledding with their kids in Central Park
and rolled up snowmen who beheld nothing.
Tickled, Ray twitted the horsebirds,
'Where's the heat?' Like a gyroscope strung
between its solar and its lunar parents,
the thawing and reclumping planet Earth
wobbled her magnetic top toward London
and her runny bottom toward New Zealand.

Eleven

Sloshing her swollen amniotic oceans
and sucking her streams into her limestone cavities,
Earth squinted through her Appalachian range
for a glimpse of the sun. Shielding her slopes
with her broad-leaved forests and shuddering all over
Oklahoma, she shook up Mexico City,
sank back as though building for another upheaval
and groaned with a cavernous vocal fry:
'Why don't you just electrocute me,
High Voltage, if you're up there! Then I won't
have to watch my own holocaust. My springs
are clogged, so fucked over are my groundwater
bowels by fracking hoses. Take
a good look. My moist and curly
Amazon crotch is plucked and shaved
and sown in soybeans and grassy beef-feed.

My Kentucky coal-hills are lobotomized,
and coal-ash ponds are blinding my lakes
and coating my cold freshwater tongues.
My seas are bowls of plastic soup.
There's an oil rush for my balding pate,
and my fuming rump is calving prematurely.
If Atlas and the rest of my ice-tits are milked dry,
if my seams and slicks and farts are ripped out of me,
if I lose my tree-fur, my water bodies,
even my sky-blue sari, if my lovely seasons
bark and froth and turn on each other,
confounded chaos will grip me for good.
Much of what I most want to save
is too far gone already. For god's sake,
snatch what you can from my denatured zones
and knock that sun-flare off his high horses.'
Earth had her say and had her fill
of the blasted heat and withdrew into her
umbrageous nether parts and waited
for Mister High and Mighty. And waited.
Then she rose moistly into her Adirondacks.
The peaked sky was unusually clear.
Nothing but a sullen patch of storm-cloud
raining its ragged shade on the mountain.
Earth snatched a cloud bolt,
twirled it by her left ear for balance
and hurled the whizzing lightning fork
that ejected the stunned charioteer from the chariot,
from his plump white boy's life.
And that's how Earth quenched fire with fire.

Twelve

Like a red giant, the horsebirds bristled,
reared and bolted, snapping their traces
and splintering their yoke. Then, as though bridled
by a whistle tuned to their ears only,
they regrouped as a four-part team and glided
down the ether toward their solar stables.
But the chariot was totaled. Here the reins fluttered,
there the tongue waggled loose
from its rotating axle-tree, over

there a few spokes grinned from a fractured
wheel-rim and everywhere mangled
car parts and Mercator shreds
spun free. The body that went
by the name of Ray sailed headfirst,
trailing its robe, its slippers, and its burnt-
gold hair. Picture a comet that looks
like it's falling but never does. Ray did.
His smoking corpse speared the ice
of Lake Oahe, a reservoir on the dammed
Missouri River. The ice smoothed
over, leaving no crack. Plunging
to the bottom, Ray dented the Dakota
Access pipeline, which anointed his astonished
head with hydraulically fractured crude.
A Standing Rock Sioux nymph
spray-painted his epitaph on the concrete dam:

Here's where the son sank.
Far did he fall for so deep a fool.

Then she rejoined her sisters downstream, and
on the legendary gravesite of Sitting Bull
they stood up a row of sticks at the shoreline.
The twigs dug down and branched right up
into a midwinter flurry of cottonwood seed-fluff.

ANTHROPOLOGY/TRAGEDY/DARK ECOLOGY

CHAPTER 1

CUNCTA FLUUNT: THE FLUIDITY OF LIFE IN OVID'S METAMORPHIC WORLD

Giulia Sissa

At the end of Ovid's *Metamorphoses*, Pythagoras of Samos (*c.* 570–*c.* 490) advocates vegetarianism as a countermeasure to the liquid ontology of a metamorphic world.[1] This eloquent speech is consistent with the vegetarian *contrainte* of the poem.[2] Metempsychosis creates both morphological dissimilarity and physical consanguinity among living beings. *Humanus cruor* may run in different bodies.[3] This compels us to ask what blood is. As the vital juice infusing raw meat and human flesh; as the fluid shed in the ritual butchery of sacrifices; as a material incarnation of the migrating soul, blood deserves special attention. This bright-red substance brings together the world. But the effusion of hypothetically human blood can also be a matter of quotidian gore, exactly as it is in what Pythagoras calls 'Thyestes' feasts'.[4] Dining is a potential tragedy.

Pythagoras draws out the consequences of the very *possibility* of metamorphosis. There is a dark side to the virtual transformation of human beings into other bodies. Changeability makes the world interconnected, but also instable and uncertain. To be sure, change is generative rather than destructive, vital rather that deadly, creative of deceptively alien, seemingly-nonhuman forms rather than obvious monsters. But, precisely for these reasons, we then belong in a network of human/nonhuman/quasi-human/also-human beings. Precisely because a metamorphosis is not a *total* replacement of human identity, but an imperceptible hybridization, it creates ambiguity, confusion and, worst of all, the risk of cannibalism. Think of Io's father, facing a plump cow. Think of Tereus feeding obliviously on his own son's flesh. Think of your own reaction at Acteon's invisibility in the eyes of his own dogs. The same possibility for error is the effect of metempsychosis. Anything comestible, although apparently nonhuman, could be 'inhabited' by a human soul. Animism, and more precisely, 'anthropogenism', as Philippe Descola has argued, is the ontological truth of this kind of world.[5]

The dissolution of humanity in the liquid ontology of metamorphosis becomes a cause for perceptual insecurity. The loss of a previous shape prevents an observer from seeing through a new body, and from recognizing the person who is now dwelling in a bovine, meaty frame. The poem creates the virtuality of alimentary violence, but then pursues three intertwined strategies to avoid it. Firstly, it crafts an internal taxonomy, which opposes animals to plants. The former are all inedible. The latter can be either safe food, when they are not the outcome of a transformation; or non-food, whenever they derive from a metamorphosis. Secondly, the poem deploys a critical discourse on bloody sacrifice.[6] Finally, it ends with Pythagoras' normative intervention, which brings together all these threads – a liquid ontology, metempsychosis, metamorphosis and the imperative to abstain, here and now, from blood-shedding and meat-eating. It is because we are at risk of burdening our own digestive apparatus with something conceivably human, that we must avoid any kind of bloody food. The speech fits both the hybrid outcome of metamorphosis, and the vegetarian logic underlying the poem.

We will start from there.

Cuncta fluunt: why Pythagoras?

Pythagoras claims that 'all things are in flux' (*cuncta fluunt*). Ages, seasons, natural phenomena and the vicissitudes of history show the mobility of life. The migration of souls produces a network of kindred beings. The ecosphere is an inclusive, affective community. This may well appear to create a welcome state of affairs, until it turns into a horror story. Since changeability does not involve the annihilation of whatever undergoes change, when it is a human being that loses its shape – and this is the case in almost all the episodes of the *Metamorphoses* – then a vestige of humanity might still be lingering in a non-anthropomorphic body. New bodies are hybrids. Since human morphology has become invisible, replaced by a new appearance, the nagging suspicion that a cow might also be a woman (or a god), remains precisely that – a supposition, a source of anxiety, never a certainty. This possibility is the profound reason for vegetarianism. Because everything changes, and because 'we can' (*possumus*) change – more precisely because our 'winged souls' (*volucres animae*) 'may enter into beastly homes and even be concealed in the bodies of cattle' (*inque ferinas possumus ire domos pecudumque in corpora condi*) – we must not ingest those bodies. They may contain the souls of human beings like us, even those perhaps of our own parents, brothers, children. This is why, Pythagoras concludes, 'Let us not load our bowels with Thyestean feasts (*neve Thyestis cumulemus viscera mensis*)!'[7]

This speech is a manifesto of animism. In the ecosphere, multiple categories of beings may well look dissimilar, but they are susceptible of partaking in the same subjectivity. Far from being a fad, the attribution of a human soul to non-anthropomorphic entities reveals an intelligible 'umbrella view' of what is there. This kind of ontology, as Philippe Descola has argued, is common to numerous societies.[8] Now, since the concluding speech of the *Metamorphoses* is, precisely, a lesson in ontology, we will approach it in these terms by asking, first of all, a question about philosophical plausibility: why Pythagoras? If we are able to answer this question, we will be better able to understand the particular manner of changeability which Pythagoras presents and which, also, occurs in the poem. The same kind of liquid, productive, animistic ontology underpins both. Its side-effects are the same. The counter-measures must be the same. We will, therefore, be able to grasp the deep coherence of the entire poem.

So, why Pythagoras? To be sure, it is not surprising that Ovid should choose a Presocratic philosopher as his internal commentator. Only a pre-Platonic ontology could be reconciled with the *Metamorphoses*. Specifically, the poem is compatible with a conception of *ta onta* τὰ ὄντα – an ontology – that merges being and time in a metaphorical language of fluency. This language has a Greek provenance. It evokes Heraclitus. In her account of Pythagoras' speech, Elaine Fantham is right in noticing this unequivocal resonance. 'Nothing in the whole world abides, but everything flows (here Ovid echoes Heracleitus)', she writes.[9] Charles Khan also sees 'Heraclitean themes, enriched with poetic echoes of Lucretius' Epicurean epic'.[10] More precisely, when Pythagoras utters the expression '*cuncta fluunt*', he directs us to look, first of all, at the motto 'all things move and nothing stays still' (πάντα χωρεῖ καὶ οὐδὲν μένει), a statement that, in Plato's *Cratylus*, prepares the suggestive comparison of the universe with a river. We will also remember that 'all things are in motion like streams' (οἷον ῥεύματα κινεῖσθαι τὰ πάντα), a claim Socrates strives to refute in the *Theaetetus*.[11] It is to Heraclitus that Socrates attributes the original assertion that all that is, *ta panta*, is in a state of perpetual flux: sensations change from one minute to the other; one cannot say anything, because words take on different meanings. One cannot even use the verb 'to be'.

> It is out of movement and motion and mixture with one another that all those things become which we wrongly say 'are' — because nothing ever is, but is always becoming.[12]

Except for Parmenides, Socrates goes on to say, philosophers such as Heraclitus, Empedocles and Protagoras have endorsed this line of thought. As poets, Epicharmus as well as Homer have shared the same idea. The *Iliad* mentions 'Oceanus the origin of the gods, and Tethys their mother'.[13] All things are the offspring of flow (*Rhea*) and motion. Now, all this moving moisture, Socrates objects, is soft, swampy thought, worthy of a frog. Not-everything flows, he counterargues. Only what falls under the senses does. There *is* 'something', stable and graspable. One may become white, of course, but there *is* 'the white'. Language itself says as much. There are transferable adjectives, but there is also the neuter. On this account, Heraclitus is credited with a theory of *radical* change, involving a plurality of things being transformed into one another. Becoming is, for him, the way things undergo constant change and incessant remixing. As Diogenes Laertius put it, in Heraclitus' own rendition of this formula, 'all things come to be according to opposition of contraries, and the whole universe flows like a river' (γίνεσθαί τε πάντα κατ' ἐναντιότητα καὶ ῥεῖν τὰ ὅλα ποταμοῦ δίκην).[14]

While reading the *Metamorphoses*, our dilemma, therefore, is the following: on the one hand, Ovid, who very probably knew of Heraclitus' ideas at least through Lucretius, might have thought that this very particular metaphor – *cuncta fluunt* — derived precisely from Heraclitus' ontology (πάντα χωρεῖ / ῥεῖν τὰ ὅλα), rather than from Pythagoras' own teachings. But, on the other hand, Ovid still chose Pythagoras. Why should Pythagoras, rather than Heraclitus, speak as the preferred theorist of an ever-changing nature, in the possible world of the *Metamorphoses*?[15]

Let us compare the beginning and the ending of the poem. Once the *Metamorphoses* set out to tell stories about *nova corpora*, new bodies, an unequivocally *vital* version of becoming is needed.[16] Becoming is coming into existence. From its first lines, the poem relies on the positive and generative resources of change. At the end, because change was meant all along to capture the way things come to be in the world, Pythagoras' metaphor of a stream – *cuncta fluunt*, all things are in flux – reinforces the poem's vibrant, energetic, creative narration about those *nova corpora*. Pointedly, Pythagoras himself speaks of nature as *rerum novatrix*.[17] Now, if the project of metamorphosis is innovation, renewal and novelty, then Heraclitus fails to fit the bill.

In Roman philosophical circles, he was known as an abstruse thinker who placed fire – a destructive phenomenon – and not water at the origin of a cosmos in constant transformation.[18] More to the point, in *De Rerum Natura,* Lucretius had aggressively criticized what he called his 'insanity', on account of the nihilism that his views on cosmic changeability implied. We know the importance of Lucretius' poem in the background of the *Metamorphoses*.[19] We can reasonably speculate that Lucretius' criticism might be the filter through which Ovid understood Heraclitus. Let us heed to Lucretius' line of thought. If we admit that every single thing is always continuously changing into something else, nothing can ever happen except via a continuous process of annihilation. There is, in the world, only the death of one thing after the other.

> For whenever a thing changes and passes out of its own limits, straightway this is the death of that which was before (*continuo hos mors est illius quod fuit ante*). Indeed, something must be left unscathed by those fires (*aliquid superare necesse est incolume ollis*), lest you find all things returning utterly to nothing (*ad nihilo*), and all things born again and growing strong out of nothing (*de nihiloque*).[20]

Lucretius insists that restless mutability implies nothingness. Continuous change brings about the logical consequence that things must come from nothing, and must end in nothing. But this, for Lucretius, is wrong. This is his damning argument *against* Heraclitus. Now, in order to rebut these faulty theories, Lucretius claims that there must be 'something' that remains 'unharmed' by fire (*aliquid* … *incolume*), 'something immutable' (*immutabile quiddam*).[21] This 'something' is the atom. Atoms aggregate and disaggregate, but never die. Atomism forestalls nothingness.

In a materialistic mode, this argument resonates with Plato's own objections to Heraclitus. In Socrates' words, Heraclitus' view of change points, once again, towards nothingness: πάντα χωρεῖ καὶ οὐδὲν μένει, 'all things move and nothing stays'.[22] The really significant consequence of flux is negative: if *all* is in a state of perpetual mobility, then *nothing* can ever stay there. And this is inadmissible. Such a relentless becoming, meant as pure nonbeing in motion, fits the comic caricature of Heraclitus' metaphysics in the *Theaetetus*. If all the 'liquid' followers of Heraclitus, including Empedocles, were to respect their own premises, they could not possibly speak. They could not even attribute qualities to objects, for instance colours to bodies, because while they are uttering their sentences, the total translation of all things would carry away what they are trying to say.[23] If their flow flows all the time, then nothing at all can ever be there for them, in any respect whatsoever. Not even language is saved. According to Plato, the 'something' that has to subsist is not the atom, as we know, but the Form. Now, we can say with a reasonable degree of certainty that Ovid must have known Heraclitus, as the nihilistic philosopher portrayed in *De Rerum Natura*. We can cautiously conjecture that he might also have known Heraclitus' thought through a Platonic filter.[24]

Now, Ovid could not possibly endorse the plausibility of this notion of change – unremitting and speechless – and certainly not in this poem. On the contrary, metamorphosis gives shape to objects and living beings, which are going to be there, in the world. They are there to stay. The poem immortalizes their becoming what they are and what they are going to be for a while. In its content-form (the ideas signified through denotation and connotation), in its expression-form (the vocabulary, the syntax, the meter), but also in the phonetic playfulness that constantly draws attention to the text itself, the poem endeavours to capture the subtlest nuances of each transformation, to seize as many instants as possible, and to track down incremental, granular, corpuscular alterations.[25] The process of change comes to life in poetic language. This is Ovid's own aesthetic and narrative thinking at work.

As a consequence, although Heraclitus' vivid image that everything is in a state of flux is attuned to the *Metamorphoses*, the poem deploys a distinctive ontology. While echoing πάντα χωρεῖ, Ovid has a different idea of such a stream. All things are in flux, *cuncta fluunt*, yes, but so as to generate *nova corpora*. The nothingness that Plato, followed by Lucretius, mockingly projects onto Heraclitus' mottos – πάντα χωρεῖ <u>καὶ οὐδὲν μένει</u>; ῥεῖν τὰ ὅλα – is gone. The intermittence of death is also gone. Hence the choice of Pythagoras. Ovid has Pythagoras utter an axiom previously attributed to Heraclitus, but with an innovative, vitalistic twist. With a little help from Pythagoras, the poet reorients fluidity toward life – life in motion through metamorphosis and metempsychosis. The touchstone of this novel orientation is a concern for the protection of living beings that, notwithstanding their unhuman body, might still be a bit human. Vegetarianism can only make sense in this context.[26]

To conclude our ontological excursus: Ovid's Pythagoras may well echo Heraclitus, but he is not Heraclitus. He also echoes Empedocles, but he is no Empedocles either. Philip Hardie has argued that Pythagoras' speech at the end of the *Metamorphoses* conveys Empedoclean insights.[27] I would like to respond with three considerations that contribute to our discussion of the poem's ontology. Firstly, the doxographic tradition attributes to Pythagoras precisely the line of thought that Ovid lends to his own Pythagoras – a soul moves from one living being to another.[28] It does not seem necessary to suppose that Pythagoras' own ideas should have been lifted from his follower, Empedocles, and then replaced in his own mouth.[29] Secondly, Empedocles' fragments convey a vision of change that, once again, may include nothingness. On the one hand, no wise man would claim that 'before mortals have coalesced and [after] having dissolved, they are nothing' (οὐδὲν ἄρ' εἰσίν).[30] But, on the other, the poet announces that 'there is birth of nothing, among all mortal things, nor is there an ending of baleful death (οὐδέ τις οὐλομένου θανάτοιο τελευτή), but only mixture and exchange of things mixed exist, and "birth" is a name given by mortal humans.'[31] This is substantially different from Pythagoras' words about birth, death and changeability, in Ovid's poem.

> Nothing retains its own form; but Nature, the renewer of things (*rerumque novatrix*), ever makes up forms from other forms (*reparat natura figuras*). Do believe me: nothing perishes in the whole world (*nec perit in toto quicquam ... mundo*); it does but vary and renew its form (*sed variat faciemque novat*). What we call birth is but a beginning to be other than what one was before; and death is but cessation of a former state. Though, perchance, things may shift from there to here and here to there, still do all things in their sum total remain stable (*constant*).[32]

Remarkably, becoming is not a matter of mixture and dissolution, as in Empedocles' fragments. The all-important Empedoclean notion that body parts aggregate and disaggregate is missing in Pythagoras' speech. Rather, forms are renewed and refurbished – upcycled, so to speak. The result is always positive: 'nothing perishes in the entire world' (*nec perit in toto quicquam ... mundo*). Death is abolished, for what we call 'death' is merely change. The emphasis is on Nature's inexhaustible creativity (*reparat natura ... novat*) in bringing new things into existence. Never are Mortals nothing – this is the point.

Thirdly, Lucretius makes precisely this point in *De Rerum Natura*. Empedocles' conception of change, he argues, presupposes nothingness. Empedocles deserves, therefore, the same objection directed to Heraclitus. According to Lucretius, there 'must' exist (*necesse est*) something that, notwithstanding the twists and turns of generation and destruction, stays there. There must be a substratum that remains intact, while being subject to motion and change, combination and dissolution. But this persisting reality cannot be made of air, fire, earth and water, because in their reciprocal transformations, these so-called primordial components of the physical world must inevitably go to nothing. Empedocles cannot be right. The permanent being that lies behind and underneath all existing bodies is rather made of atoms and void. From Lucretius' point of view, by ignoring atoms and void, Empedocles reveals himself to be as much a nihilist as Heraclitus is.

> For there must be something that abides unchangeable (*immutabile enim quiddam superare necesse est*), that all things be not altogether brought to naught (*ad nihilum*). For whenever a thing changes and passes out of its own limits, straightway this is the death of that which was before (*continuo hoc mors est illius quod fuit ante*). Therefore, since these things (the four elements) which we mentioned a little while ago pass into change, they must of necessity consist of other things which can nowhere change at all (*quae nequeant convertire usquam*), or you will find that all things return utterly to nothing (*ne tibi res redeant ad nilum funditus omnes*).[33]

We do not know if, or how well, Ovid might have known at first-hand the writings of these early Greek philosophers, but we do know that he deeply admired 'sublime Lucretius'. It is reasonable to reckon that Ovid may well have looked at Empedocles, as Philip Hardie argues, through a Lucretian prism.[34] I concur. But that prism was so deeply critical that Lucretius' criticism makes Empedocles' ontology especially unsuitable for the *Metamorphoses*. A 'reading' of Empedocles via Lucretius would have encouraged

Ovid to disagree, rather than agree, with a thinker accused of nihilism. My hypothesis is that Ovid might have known Empedocles' thought not merely via Lucretius, but in the wake of Lucretius' *refutation* of nothingness. With his virtually teratogenic aggregation and disaggregation of limbs, with his views on becoming, Empedocles could hardly be either an exemplary model for the *Metamorphoses*, or a philosophical authority in the poem. Which explains, once again, his absence and the presence of Pythagoras whose ideas, as far as we know, do not clash with Ovid's own ontology.

To conclude: in *De rerum natura,* both Heraclitus and Empedocles are a target of blame. Pythagoras is non-existent. In the *Metamorphoses*, neither Heraclitus nor Empedocles is featured. Pythagoras is, and prominently so. Through a Lucretian lens, the former could hardly be credited with the vitalistic ontology that informs the *Metamorphoses*. Ovid attributes to the latter a consonant philosophical voice. Pythagoras explains that our own soul, or that of a neighbour, can be reincarnated in a bovid. The poem narrates that, in the anatomy and physiology of a heifer, Io is still alive. Same ontology. Same animism. The world is incessantly renewed, repaired, diversified. We are all interconnected. Happy together?

Our mouth, our bowels

This is the nagging question from which we have started. Now, it is time to understand how we can inhabit a metamorphic world, either euphorically or uncomfortably.

Animals and plants are not only objects of knowledge and contemplation for us. We are used to feed on them. What if we were to eat foodstuff derived from metamorphosis? Would not we commit cannibalism? Perhaps, Ovid's possible world is not to be enjoyed, but to be experienced with epistemic and, worst, alimentary qualms. I have already quoted a crucial passage: 'we can' (*possumus*) be reborn as novel bodies; our human 'winged souls' (*volucres animae*) 'may enter into beastly homes and even be concealed in the bodies of cattle' (*inque ferinas possumus ire domos pecudumque in corpora condi*). As a consequence, we must abstain from those bodies. 'Let us not fill up our bowels with Thyestean feasts (*neve Thyestis cumulemus viscera mensis*)!'[35] It is on this *virtual* kinship of humans and nonhumans that Pythagoras grounds his exhortation to vegetarianism. In a world where metempsychosis is possible, a non-human animal *might* be, also, human. You shall not eat their flesh! This is what Thyestes did inadvertently when his twin brother, Atreus, served him the flesh of his own sons, or what the Cyclops used to do as a matter of course.[36] Abstention from meat is a logical – and here the modality is 'necessity' – antidote against the logical consequence of blurring taxonomic distinctions. It is an ontological precaution – what if this cow, haplessly wandering out there, might still be also Io, or Zeus. In doubt, according to Pythagoras, we must decline any serving of stew – we might be helping ourselves to a piece of Io's soft tissues. Vegetarianism is a form of alimentary scepticism.

Once again, it is humanity that matters. In the possible world of Ovid's *Metamorphoses*, we, humans, incur multiple dangers: of being transformed, of not being able to detect a transformation and, finally, of slaying, carving, cooking, chewing and then burying in our own guts a possibly human hybrid, even perhaps a child of ours – flesh of our flesh. Morphological vulnerability causes perceptual insecurity and potential omophagy – for us. Because cannibalism is possible, then vegetarianism becomes necessary – to us. Lest we renounce any meat, we are at risk of ingesting human flesh, of polluting our mouth and stomach in gory banquets. This is no light matter. The modal logic of possibilities and necessities becomes an experience of anguish, fear and doubt. Any mouthful of veal, pork, beef or goose might turn out to be anthropophagy. Thyestes is the paradigm of carnivorous eating-habits. Tragedy lurks.

Human blood

I would like to focus on a particular fluid that runs through this possible world: blood.

Blood is a challenging substance. Blood divides – bloody *versus* bloodless beings. Red *versus* green. It is an operator of taxonomy. But, if metamorphosis is possible, then blood must or can 'circulate', so to speak, from human bodies to other bodies. This fluidity becomes a terrifying menace. According to Pythagoras, although it is the soul (*animus*) that can move from a person to another living being, it is the matter of the body that we must refuse to consume and, even more precisely, it is *human* blood, *humanus cruor*, that we are in great danger of tasting.[37] In an especially poignant passage, while projecting human sensibility onto the animals we may slaughter for alimentary consumption, Pythagoras laments that our proximity to these creatures is not merely a matter of similar behaviour: it is a form of potential consanguinity. Human souls, we said, can be concealed (*condi*) into a different animal. And here Pythagoras asks his passionate questions: 'Who is capable of shedding "human blood", *humanus cruor*, by slaying them?' 'What is the distance between this and full-fledged murder (*plenus facinus*)?' Hence the normative conclusion: we are permitted to kill the animals that threaten us like enemies at war, but we must *never* eat their flesh! More dramatically: 'let our mouths be empty of meats and let them take in meek foods (*ora vacant epulis alimentaque mitia carpant*)!'[38] Our mouth (*ora*) and, as in the passage quoted earlier, our bowels (*viscera*) – this is what matters. Physical contact. Incorporation. To ingest a probably hybrid flesh is to continue the metamorphic 'concealment' of a human being into a new body. If Io hides in a cow, by consuming a steak, we would be intaking her material substance – her meat, and the juice that imbues that tissue, *cruor* – in our own digestive apparatus. The transfer of souls actually coincides with a 'transfusion' of human blood. Transmigration is transubstantiation. This is the logic of vegetarianism.

Now, a question must be raised: is blood the material substance that embodies the identity of a living being?[39] Cicero had mentioned a similar dilemma, whether the soul might be fire, air or blood (*utrum sit ignis an anima an sanguis*).[40] Now, according to Cicero, the source of the latter theory, *animus* equals *sanguis*, was a philosopher: 'Empedocles believes that the soul is blood impregnating the heart (*Empedocles animum esse censet cordi suffusum sanguinem*)'.[41] Lucretius, on the contrary, had accused ordinary people, not a master of truth, of entertaining this belief, namely of thinking 'that they know that the nature of the soul is blood or even wind (*se scire animi naturam sanguinis esse aut etiam uenti*).'[42] Another point of disagreement with the masses, and, if Cicero is right, with Empedocles! In Greek culture, the same identification of the soul with blood transpires in an especially authoritative account of the body, the *Hippocratic corpus*. A medical writer criticizes those physicians who reduce the nature of a human being to one substance only, for instance blood. 'Those too who say that man is blood (οἱ τὸ αἷμα φάντες εἶναι τὸν ἄνθρωπον) use the same line of thought. They see men who are cut bleeding from the body, and so they think that the soul of a man is blood (τοῦτο νομίζουσιν εἶναι τὴν ψυχὴν τῷ ἀνθρώπῳ).'[43] In Plato's *Phaedo*, 96 b, Socrates mentions, among the questions he used to ask when he studied nature, the dilemmas concerning the soul: 'Is it the blood, or air, or fire by which we think (καὶ πότερον τὸ αἷμά ἐστιν ᾧ φρονοῦμεν, ἢ ὁ ἀὴρ ἢ τὸ πῦρ)?'[44] Again, Aristotle mentions such a theory, without endorsement, in his *De anima*.[45]

How about Pythagoras himself? In Diogenes Laertius' account, the soul extends from the heart (where spiritedness is located) to the brain (the seat of reason and intelligence). Whereas nonhuman animals are devoid of reason, they share intelligence and spiritedness with humans.[46] Only reason is immortal, all else is mortal. The soul is not blood, but 'it is nourished from blood (τρέφεσθαί τε τὴν ψυχὴν ἀπὸ τοῦ αἵματος)'. In Ovid's poem, Pythagoras speaks as if *humanus cruor* might flow, hypothetically, in *any* animal body. Meat, any meat, as a consequence, is off-limits. The *alimenta mitia* that are placed in a binary opposition with this cannibalistic food can only be vegetables. The poem celebrates them. Like Pomona, Baucis and Philemon, the Roman readers can freely help themselves to apples, berries, figs, grapes, honey and plenty of common vegetables.[47] Like their blissful ancestors, living in the Golden Age, they can gather arbutus fruits, wild strawberries, and blackberries; they can collect 'the sweet acorns fallen on the ground from Jupiter's large tree (*et quae deciderant patula Iovis arbore glandes*).'[48] Nutritious cereals have been available since the silver age.[49] This mythologized periodization of natural history agrees with Pythagoras' own appeal to remember a pre-carnivorous past: once upon a time, again in that age we call 'golden', fruit and herbs were our only diet, and we 'never would pollute the mouth with blood'. Birds, fish and feeble animals were not afraid of us. 'All things were safe (*tuta*) from insidious wiles, fearing no deception. All things were filled with peace (*cuncta sine insidiis nullamque timentia fraudem plenaque pacis erant*).'[50]

Green peace

Pax, peace: no hunting, no sacrifice, no inter-species war. Again, war is the model of violence against non-human living beings. Vegetarianism is pacifism.

In this taxonomic logic, therefore, vegetarianism shields us from feeding on blood as a potentially 'trans-human', homogenizing stuff, a vestige of human flesh gorging other kinds of flesh – what in English is reassuringly called 'meat'. Vegetarianism forestall vampirism, so to speak. But how can we be sure that plants are bloodless? Wherever metamorphosis is possible, human blood (*humanus cruor*) may always surface and be shed. Since there exists vegetation of human provenance in the poem, how about these potentially hybrid tissues: should they not bleed?

A number of vegetables prolong the human life of a pre-existing creature, now 'engrafted' in their trunks, stalks, shafts, branches and blossoms. These plants may ooze a variety of bodily fluids. Once made into poplars, the daughters of the Sun, the Heliades, never stop mourning their brother fallen from the sky, Phaethon. They are always weeping.[51] The lacrimous tree in which Myrrha, an incestuous daughter, is now alive – and prey to eternal regret – keeps leaking a precious perfume. 'She cries however, and lukewarm droplets trickle from the tree. There is honour to the tears, and a stillicidium of myrrh flows from the hard-wood (*flet tamen, et tepidae manant ex arbore guttae. Est honor et lacrimis, stillataque ex robore murra*).'[52]

Blood too is involved in metamorphosis. Dryope, plucks some lotus flowers for her small child to play with. Drops of blood fall from the fractured blossom (*guttas e flore cruentas decidere*).[53] The branches move in trembling horror (*tremulo ramos horrore moveri*).[54] And then, in slow motion, the lotus starts to incorporate Dryope herself, who now becomes a tree. When she maimed the plant, Dryope was not aware that the lotus was also Lotis, a young woman pursued by Priapus' obscene assaults. Erysichthon is an arrogant man who fails to pay respect to a monumental oak, sacred to Demeter. The goddess herself lies under the tree. He attacks the oak with an axe. The tree moans and grows pale. Blood gushes from the wound (*vulnus*), as if it were a sacrificial victim.

> Blood flowed from the shattered bark, not differently from the blood that is usually shed from a broken neck, in front of the altars where, as a large victim, a bull falls to earth.[55]

These sad stories offer elaborate variations on a pathetic theme, that of the wounded, lamenting, weeping and, more disturbingly, bleeding plant.[56] A metamorphic world becomes a web of consanguineous beings. Blood is shared and shed. Which raises the question whether there exist any pure 'greens'. The answer is 'Yes'. If we read carefully, we discover that in the poem's vast vegetal realm, only certain kinds of plants derive from metamorphosis. These are floral or arboreal beings, such as the laurel, the lotus, the poplar, the oak and the myrrh tree. Having a human provenance, they must be as incompatible with alimentary consumption as meat is. Luckily, they are inedible. In contrast, there exist a profusion of other greens, cereals, roots, and fruits that, in the twists and turns of the narrative, seem to have escaped metamorphic hybridization. These are bloodless and, as a consequence, edible food. Think of Pomona's orchard, rife with succulent fruit, of Baucis' earthy cabbage, and of the simple fare growing spontaneously in the fields in the Golden Age.[57] All these veggies are good to eat, both because they taste delicious and because they are safe. Having always been what they are, they cannot conceal a lingering human presence.

Not only is the biosphere of the poem wonderfully organized, but the distinction of edible *versus* inedible plants fits perfectly the logic of Pythagoras' vegetarianism. No foodstuff should be traceable to human origins, lest we might fall into involuntary cannibalism. How can we survive? Whereas one food-group is systematically treacherous, because *any* seemingly nonhuman animal might be unrecognizably human, plants must necessarily be the alternative – and fortunately, many of them are animal-free, thus human-free. The solidarity between Pythagoras and the poem hinges on this logic. Pythagoras teaches that we *must* limit our regimen to this kind of uncontaminated produce. We *must* eat only apples, berries, honey, milk. The poem certifies, so to speak, that none of these horticultural delights was ever human. The poem abides by this *contrainte*, so that vegetarianism becomes both necessary (in order to avoid all meaty food) and realistic (plenty of plants are edible and peaceful, *alimenta mitia*). Ovid masters the logic of his possible world to perfection. Nature and cuisine conform to a taxonomic order. They fit an overarching narrative that presupposes an environment in which everything has its place.

Water too has its place. When a human being changes into a spring, blood may become drinkable water. But the poem makes sure that water is not discoloured blood. Abiding by the same logic, when water has to be potable, the poem takes the precaution of *replacing* blood with water. On the one hand, Myrrha's own 'blood goes into sap' – *sanguis it in sucos*.[58] On the other, Cyane, a nymph who tries to stop the raptor of Proserpina, cries and melts into water. Now the bones can be flexed, the nails lose their rigidity, the blond hair, the fingers, the legs and the feet liquefy (the transition to cold waves is short for limbs that are thin); and then the back and the sides and the chest flow away in light rivulets. And finally, the poet adds, 'instead of lively blood' (*pro vivo sanguine*), water (*lympha*) runs in her maimed veins – and nothing remains that you could possibly grasp.[59] Water is *not* blood.

Victory on the victim

Carnivorous habits are engrained not only in ordinary customs, but in rituals involving the gods. Animals are slaughtered, partially offered to the gods and eaten among the sacrificers. Especially under Augustus' direction, these practices were essential to Roman religious and political life. The Altar of Augustan Peace, the *Ara Pacis Augustae*, displays a monumental scene of sacrifice. The immolation of animals can be advertised in an unabashed, proud and triumphal fashion. In light of this unqualified celebration, Ovid's numerous attempts to expose the gore and cruelty of ritual slaughtering convey an ambivalent message, to say the least. If there has ever been a period of peace, this was the Golden Age, a blissful time when nobody would have dreamt of killing a living being, possibly a fellow human or a dear one.

Ovid's dissident discourse, based on the neat distinction of plants (edible/inedible) versus animals appears to be even more striking, once we place it against the background of Roman sacrificial practice. As John Scheid has argued, 'animal sacrifice ritually displays the social order and hierarchy in this world (gods – humans – animals). The offering of plants adds a complementary figure to this representation, by enlarging the hierarchy of being.'[60] Scheid has pushed the argument to the radical conclusion that, in rituals actually performed, there was no difference between meaty and vegetal offerings. The sacrificial procedures were the same; plants were always part of bloody sacrifices. However, 'nothing prevents scholars, poets and antiquarians from interpreting the double nature of sacrificial offerings, by distinguishing them rigorously'.[61] Now, Ovid is precisely one of those learned poets and antiquarians. He does draw a line between bloody and bloodless rituals. In the *Metamorphoses*, King Numa, Pythagoras claims, in order to pacify the belligerent Romans, introduced sacrificial rites, (*sacrificos docuit ritus*).[62] Pythagoras fails to describe these 'sacrificial rites', but, according to Plutarch, they must involve the offering of flour, wine, and simple gifts with no effusion of blood.[63] In contrast, Pythagoras emphasizes the absurd violence of slaying domestic animals to the gods, as if the gods could possibly welcome that brutality. *Metamorphoses* deviates from the taxonomic norm.

While reading the poem itself, Andrew Feldherr has fully acknowledged Pythagoras' concern, first and foremost, with sacrifice.

> Pythagoras' speech has been treated (and dismissed) primarily as an exhortation to vegetarianism, but almost from the beginning of his tirade, the argument against eating animals becomes a condemnation of sacrifice.[64]

Andrew Feldherr goes on to argue that Pythagoras' depiction of a heifer looking at the preparations of her own immolation resonates with the preparations of a sacrifice to Jupiter in *Metamorphoses*, 3, 10–26, and that the entire speech intimates at the finitude of the Roman empire, casting a shadow on Augustan glory. Feldherr raises a crucial question:

> Does Ovid, like his character Pythagoras or his predecessor Lucretius, generate sympathy for the victim in order to indict the practice of sacrifice itself, and by implication the visions of society and the cosmos that sacrificial ritual constructs?[65]

The answer is that 'Ovid's interest in vividly reproducing the victim's perspective can be understood as an effort not so much to undermine sacrificial practice as to appropriate it'.[66] I am afraid, I cannot agree. The poem does 'undermine sacrificial practice'. The 'condemnation' of the ritual slaughtering of animals runs deep through the *Metamorphoses*.[67] On the one hand, the poem projects into the past a blissful condition in which nobody ever shed blood either to stay alive or for the sake of the gods. We could not possibly doubt that the Golden Age, as a 'green' age, is praiseworthy. On the other hand, in the Trojan-Graeco-Roman universe of the poem, animal sacrifices do take place as it is to be expected in these cultures, but the manner of framing these scenes, of narrating the gestures, and of describing the details betrays a judgement, a blame.

Sometimes, the rituals are purely and simply aborted. Philemon and Baucis want to offer a goose to their divine visitors, Jupiter and Mercury, but the bird is never slaughtered. The two gods themselves refuse the offer.[68] Similarly, in the history of the foundation of

Thebes, the sacrifice of a heifer to Jupiter is never carried out.[69] Often, the ceremonies take place in the most murderous possible way. In Book 12, Iphigenia is presented as a victim to Diana, but at the very last minute the goddess transforms the young woman into in a doe. The doe is killed. The killing is called 'murder' (*caedes*), which is consistent with the hybridization of a woman now changed into a new body. This is not a replacement; Iphigenia is still there. Cruel 'Diana is appeased by this murder (*lenita est caede Diana*)'.[70] In Book 13, the slaying of Polyxena is fully accomplished.[71] And it is terrible. In any sacrificial situation, blood is shed, and this *cruor* might be – and, if metamorphosis is possible, could be – that of a human. The interchangeability of human and nonhuman victims underscores the invisible circulation of our own precious *human* life among other living beings – what Pythagoras warns us about. Life is embodied as blood. Let us keep that shared gore away from our lips! Because we are consanguineous, we could easily become cannibals. While representing bloodshed in this light, Ovid creates an atmosphere of unease around the act of killing animate beings, which resonates with Pythagoras' 'physical horror', at the risk of eating meat.[72]

We can discern a coherent discourse on sacrifice in the poem. Let us pull together the multiple threads we have been pursuing: Pythagoras' normative celebration of vegetarianism; the vegetarian *contrainte* that runs through the *Metamorphoses*, namely the neat distinction between animals (all inedible) and plants (not all inedible). Now, if we connect these structural features with the theme of sacrifice, a line of thought comes to light. Both Ovid's and Pythagoras' voices contribute to an overall dissident perspective on banalized blood-shedding. The most anodyne religious custom is deeply disturbing. With a didactic nudge from Pythagoras, the narrative itself calls into question Roman mores.

Like Andrew Feldherr, Dennis Feeney has argued that Ovid casts a provocative doubt on the ritual. Whereas Virgil's *Georgics* presents us with a 'range of possibilities', so that 'animal sacrifice is open to multiple interpretations', Ovid 'represents sacrifice as a token of the loss of the Golden Age'.[73] This is a consequential statement, that I wish to reinforce. If it is correct, and I think it is, it means that Ovid looks at the most official of Roman rites as a routine of glorified violence. Such a violence runs against the behaviours and the values of the only blissful era that, according to the *Metamorphoses*, ever existed in the history of human kind. This was a time when 'all things were filled with peace' (*cuncta . . . plenaque pacis*). Sacrifice, and the victorious triumph over a *victima*, a defeated 'loser', is the very opposite of that peace. The present is a state of war. More systematically, I will add, the ritual immolation of animals reveals that some individuals show a complete indifference to the foundational premise of their own ecosystem. The narrative universe of the poem is haunted, as we have emphasized, by the possibility of metamorphosis.

Narration is paramount. Whereas in Xenophanes (via Diogenes Laertius), Pythagoras is able to recognize at once the voice of a dear soul trapped of in a puppy's body, Ovid's Pythagoras insists on people's insensitivity.[74] Pity has to be learned. It is the poem itself that – from the beginning to its didactic final – teaches it, by recounting the vicissitudes of hybridization. The entire poem enlightens us, the readers. We learn about the possibility of an undetectable human presence in nonhuman beings. We learn that our death is someone else's life. Here and now, any of us can (*possumus*) be suddenly reduced to a mine of juicy red meat – by dying naturally and returning to life reincarnated in a comestible beast. What happened to Io could happen to me, simply because I am mortal. I could end up being either mincemeat, or a flesh-eater. At odds with this anxiety, the ritualistic habit of slaying a victim to a divinity supposes that it does not matter in the least whether a cow might be, also, a young woman whose blood could contaminate our own mouths, lips and bowels. The Romans are indeed cannibals. It is in their daily life that they enjoy Thyestes' meals, as a matter of course. The Prince is happy. No tragedy in view. Who cares?

Bloody sacrifice in the *Fasti*: war and peace

Once read in its entirety, the *Metamorphoses* offers a manner of eco-criticism. The liquid ontology of *cuncta fluunt* brings together all the components of the world. Everything is under the sway of metamorphosis. Distinctions are undone. But the problem is that distinctions help us know what we are doing. It is not that the Roman world is chaos. It is that anything could also be something else, at the same time. This ontology contrives a systemic, ironic subversion of Augustan culture. Precisely at the moment when the poet seems to celebrate the *princeps*, in Book 15, Pythagoras undermines the religious foundation of his authority.

Even more provocatively than in the *Metamorphoses*, the same logic transpires in the *Fasti*. In an enlightening discussion of Ovid's critical take on sacrifice, Stephen Greene asks two interrelated questions.[75] Firstly, he probes Ovid's overall vision of bloody sacrifice. Both in the *Metamorphoses* and in the *Fasti*, Greene argues, Ovid is ambivalent. Descriptions of sacrifices as if they were normally performed, are compatible with a persistent Pythagorean overtone. Secondly, Greene wonders how we can reconcile such an ambivalence with what seems to be, in both poems, a praise of Augustus. The poet's fluctuation about ritual blood- shedding acquires a political significance. At the end of Book 1 of the *Fasti*, the narrator guides us to contemplate a grandiose monument, the *Ara Pacis Augustae*. Let us read along with Stephen Green.

> Perhaps the most positive picture of animal sacrifice comes at the end of Book 1, when Ovid, in the guise of Master of Ceremony, orchestrates the religious proceedings at the Altar of Augustan Peace (1.719–22): *tura, sacerdotes, Pacalibus addite flammis, albaque perfusa victima fronte cadat; utque domus, quae praestat eam, cum pace perennet ad pia propensos vota rogate deos*. 'Add incense, o priests, to the Peaceful flames, and let a white victim fall with drenched forehead, and ask the gods, who are favourably disposed to pious prayers, that the house which assumes responsibility for peace live forever with peace.' In this ceremony, every process is correct. [. . .] Augustus could have asked for no greater promotion of the connection between animal sacrifice, imperial family, and Roman prosperity than this.[76]

A white victim falls to the ground, and *now* there will be peace for ever after. Really? We have seen that, in the *Metamorphoses*, if there had ever been a time in which all things were 'full of peace' (*plena pacis*), that would have been the Golden Age.[77] No hunting, no sacrifice: *that* was peace. But here and now, on the bas- reliefs of the Altar of Augustan Peace (*ara pacis augustae*), the word *pax* resonates politically. The sculptures associate the immolation of a white victim, *alba victima,* with the perennial peace brought to fruition by the house of the Prince. While detailing that scene, the poet's voice tells us that killing animals is actually a kind of war. 'The *vict*im (*victima*) is so called because it falls under a *vict*orious right hand; the sacrificial victim takes its name from conquered foes (*vict*ima, quae dextra cecidit *vict*rice, vocatur; hostibus a domitis hostia nomen habet*).'[78] A captured loser falls to its knees, not by itself, but under the heavy hand of a winner: this is the truth of bloody sacrifice, not the *pax augusta*. It is a 'war' between humans and animals that lies at the core of the ritual.[79] This war is not even just.

It is along these lines that the *Fasti* offers a long aetiological account of the first bloody sacrifices.[80] Once upon a time, a goat and a pig attacked Ceres and Liber. This is why humans started to kill them in an act of righteous defence. But the ox and the lamb, the poet goes on to say, never did anything wrong.[81] Nevertheless, we slay them all. Animals that 'deserved' to be punished for an attack on a divinity may well call for a counter-attack. But many others, such as the ox, the lamb, the doe, the horse, the donkey, the dove, the goose and the rooster never waged any war on anyone. Nevertheless, we slay them all. We do so on behalf of multiple gods – Ceres, Priapus, Liber, or Venus. But it is we who initiate the carnage. Only occasionally, the poet concedes, can sacrifice be considered a sort of just war. Most of the time, by laying hand on our *victimae*, we exercise our belligerent domination, our *imperium* over innocent living beings. This is the unvarnished truth of animal sacrifice. The attempts to put the blame on the victims are wrong. 'The sow suffered for her crime, and the she-goat suffered, too, for hers. But the ox and you, placid sheep, what did you commit (*culpa sui nocuit, nocuit quoque culpa capellae: quid bos, quid placidae commeruistis oves*)?'[82] If *all* animals can be condemned to death, including the most harmless and generous helpers, then *no animal* can ever be protected. 'What remains safe, when even the wool-bearing sheep and ploughing oxen lay down their souls upon the altars (*quid tuti superest, animam cum ponat in aris lanigerumque pecus ruricolaeque boves*)?'[83] The *culpa* of a few particular creatures is merely an excuse for generalized slaughter. The violence of sacrifice is indiscriminate. Pure victory. This blanket aggression creates insecurity.

Whereas the possibility of metamorphosis haunts the world of the *Metamorphoses*, the reality of sacrifice betrays the violence inherent in the world of the *Fasti*, namely Rome.[84] The addressee of this poem, Germanicus, will recognize the innumerable *sacra* that fill the Roman calendar, day after day. 'Let others sing of Caesar's wars; let me sing of Caesar's altars (*Caesaris arma canant alii: nos Caesaris aras*).'[85] The *ara pacis augustae* is precisely one of those altars. The opening lines of the *Fasti*, therefore, set the stage for an anthropologically significant lesson. Here and now, we, Roman readers, can expect to learn about all the rites that do take place on the many *arae* of the Prince; we can expect to discover all the occasions –or all the excuses – for us to butcher yet another innocuous beast. In this world, Rome, no non-human animal can ever be safe, and we, humans, triumphantly relish one Thyestean feast after another. We have read the *Metamorphoses*. We have listened to Pythagoras. Still, *cruor humanus* is our daily fare.

Ovid has invented eco-criticism. The Prince did not fail to notice.

Notes

1. Ov. *Met*. 15.460–78. Ovid's text and translation are quoted from Ovid, *Metamorphoses*. Translated by Frank Justus Miller. Revised by G. P. Goold. Loeb Classical Library, volumes 42 and 43. Cambridge, MA: Harvard University Press, 1916. The translation has been occasionally modified. I would like to thank Francesca Martelli, Marco Formisano, Christopher Riedweg and the anonymous reviewer of our manuscript for their helpful thoughts.
2. On the speech's consistency with the poem, see Hardie 1995, Riedweg 2002: 48–51; on the vegetarian *contrainte*, Sissa 2019, with bibliography.
3. Ov. *Met*. 15.463–4.
4. Ov. *Met*. 15.459–62.
5. Descola 2005 and 2013. On the intertextual connection of Ovid's Tereus (*Met*. 6.412–74) and Seneca's *Thyestes*, see Schiesaro 2003.
6. Habinek 1990; Feldherr 1997; Prescendi 2007; Green 2008; Scheid 2012.
7. Ov. *Met*. 15.459–62.
8. Descola 2005 and 2013.
9. Fantham 2004: 115.
10. Kahn 2001: 147.
11. Pl. *Cra*. 402 a; *Tht*. 160 d. Cf. *Tht*. 180 b: πάντα κινεῖται; 181 c: 'Exactly what do they mean, after all, when they say that all things are in motion? (ποῖόν τί ποτε ἄρα λέγοντές φασι τὰ πάντα κινεῖσθαι).'
12. Pl. *Tht*. 152 d-e: ἐκ δὲ δὴ φορᾶς τε καὶ κινήσεως καὶ κράσεως πρὸς ἄλληλα γίγνεται πάντα ἃ δή φαμεν εἶναι, οὐκ ὀρθῶς προσαγορεύοντες: ἔστι μὲν γὰρ οὐδέποτ᾽ οὐδέν, ἀεὶ δὲ γίγνεται. See also 157 a-d: 'And so it results from all this, as we said in the beginning, that nothing exists as invariably one, itself by itself, but everything is always becoming in relation to something, and "being" should be altogether abolished, though we have often – and even just now – been compelled by custom and ignorance to use the word. But we ought not, the wise men say, to permit the use of "something" or "somebody's" or "mine" or "this" or "that" or any other word that implies making things stand still, but in accordance with nature we should speak of things as "becoming" and "being made" and "being destroyed" and "changing"; for anyone who by his mode of speech makes things stand still is easily refuted. And we must use such expressions in relation both to particular objects and collective designations, among which are "mankind" and "stone" and the names of every animal and class. Do these doctrines seem pleasant to you, Theaetetus, and do you find their taste agreeable?'
13. *Il*. 14.201, 302.
14. D. L. 8.8.
15. On Ovid's philosophical options, see Beasley 2012. On the reception of Heraclitus in Roman culture, see Saudelli and Lévy 2014.
16. Ov. *Met*. 1.1–3.

17. Ov. *Met.* 15.252–8.

18. Cicero, *Div.* 2.133: *valde Heraclitus obscurus.*

19. On Ovid and Lucretius, see the Introduction in Barchiesi 2005; Sissa 2010 with bibliography.

20. Lucr. *DRN* 1.670–4. See also 1.881–6, on Anaxagoras: 'if there were minute substances, hidden in material objects, then we should be able to see blood in fruit, when we crush them, or in stones when we stroke them, or milk from herbs . . . but we do not see these things. There are only atoms.' Lucretius' text and translation are quoted from *Lucreti De Rerum Natura: Libri Sex* (Second Edition), Cyril Bailey (ed.), Oxford Classical Texts, Oxford University Press, 1921. The translation is occasionally modified.

21. Lucr. *DRN* 1.672: *proinde aliquid superare necesse est incolume ollis*; *DRN* 1.790: *immutabile enim quiddam superare necesse est*. For a brilliant discussion of the precise intertext of *DRN* 1.790–3, and *Metamorphoses* 15.254–8, see Centamore 1997: 234: 'For Lucretius, in fact, change is equivalent to leaving one's boundaries, losing one's identity. The atom, therefore, is a guarantee of the immutability of the cosmos. Without the atomic components, phenomenal dynamics would fall into complete annihilation . . . Change, therefore, causes the death of the individual compound, be it thing, plant, animal or man. On this point Lucretius is categorical: "*Quod mutatur enim dissolvitur, interit ergo*" (Lucretius, *DRN* 3.756). Ovid, on the other hand, seems to oppose this conception. The Pythagoras of the *Metamorphoses*, in fact, reverses the meaning of the Lucretian verse by stating: "*Omnia mutantur, nihil interit*" (Ov. *Met.* 15.165).' Giordano Bruno relies on Pythagoras' speech, in order to save a principle of continuity through natural transformations. 'In essence, Bruno seeks to hinge the materialist model within a Platonizing conception' (pp. 239–40, my translation).

22. Pl. *Cra.* 402 a. The expression reappears, more developed, at the end of the dialogue. See 439 d – 440 e, in particular 440 c-d, about those who 'believe that all things flow like pots, and believe that just like people sick with catarrh, so all things are in such state – they are affected by rheum and catarrh (πάντα ὥσπερ κεράμια ῥεῖ, καὶ ἀτεχνῶς ὥσπερ οἱ κατάρρῳ νοσοῦντες ἄνθρωποι οὕτως οἴεσθαι καὶ τὰ πράγματα διακεῖσθαι, ὑπὸ ῥεύματός τε καὶ κατάρρου πάντα τὰ χρήματα ἔχεσθαι). For the medical provenance of these comparisons, see Hippocrates, *Aphorisms* 3.12.

23. Pl. *Tht.* 192d–183c.

24. On Ovid's knowledge of, and possible references to Plato, see Wheeler 1995. On Lucretius' sources, see Monterese 2012. See also Fratantuono 2015 (since fire is destructive for Lucretius, it cannot be creative). On Pythagoras and Plato, see Long 1948, Kahn 2001, Riedweg 2002.

25. On these useful notions, introduced by Louis Hjelmslev, after Ferdinand de Saussure, see Eco 1975, chapter 3.7.

26. D. L. 8.13; 19; 22. The motto πάντα ῥεῖ does not appeare in Diogenes' account of Pythagoras' thought.

27. Hardie 1995.

28. This is a selection of relevant passages from Diogenes Laertius: D.L. 8.1.28 ('Soul is distinct from life; it is immortal, since that from which it is detached is immortal. Living creatures are reproduced from one another by germination; there is no such thing as generation from earth'); 8.1.4–5 (Pythagoras' own reincarnations); 14 ('He was the first, they say, to declare that the soul, bound now in this creature, now in that, thus goes on a round ordained of necessity'); 8.1.30–1 ('The soul of man, he says, is divided into three parts, intelligence, reason, and spiritedness. Intelligence and spiritedness are possessed by other animals as well, but reason by man alone. The seat of the soul extends from the heart to the brain; the part of it which is in the heart is spiritedness, while the parts located in the brain are reason and intelligence. The senses are distillations from these. Reason is immortal, all else mortal. The

soul draws nourishment from the blood (τρέφεσθαί τε τὴν ψυχὴν ἀπὸ τοῦ αἵματος); the faculties of the soul are winds, for they as well as the soul are invisible, just as the aether is invisible. The veins, arteries, and sinews are the bonds of the soul. But when it is strong and settled down into itself, reasonings and deeds become its bonds. When cast out upon the earth, it wanders in the air like the body'); 8.1.33 ('Purification is by cleansing, baptism and lustration, and by keeping clean from all deaths and births and all pollution, and abstaining from meat and flesh of animals that have died, mullets, gurnards, eggs and egg-sprung animals, beans, and the other abstinences prescribed by those who perform mystic rites in the temples'); 8.1.36 (Xenophanes confirms the statement about his [Pythagoras] having been different people at different times in the elegiacs beginning: Now I show other thoughts, another path. What he says of him is as follows: They say that, passing by when a puppy was hit hard, he, full of pity, said: 'Stop striking hard! This is the soul of a dear man! I recognized that soul at her voice! (παῦσαι μηδὲ ῥάπιζ', ἐπεὶ ἦ φίλου ἀνέρος ἐστὶ ψυχή, τὴν ἔγνων φθεγξαμένης ἀΐων)'); 8.1.44. Iamb. See also *VP* 25; Porph. *Abst.* 1.23–6. On the peculiarities of transmigration according to Pythagoras, see Pellò 2018 (metempsychosis is not a punishment); Horky 2021 (the migrating soul could be long-lived rather than everlasting, material and made up of atoms).

29. On their respective ideas, see Palmer 2018. See also Primavesi 2007, Newmyer 2014, Kleczkowska 2017 on the specificity of Empedocles' vegetarianism, justified on account of the physical kinship of all living beings as such, without any explicit mention of the transmigration of individual souls. The movement of a soul from body to body, on the contrary, is Pythagoras' point in the *Metamorphoses*. Taxonomy is narrativized. The possibility of cannibalism is thematized.

30. Emp. D52 (B15), *ap.* Plut. *Adv. Col.*, Laks and Most 2016: 397.

31. Emp. D53 (B8) Aëtius. I have modified the translation by Laks and Most 2016 ('nor is there an ending coming from baleful death'), because it misses the point (p. 397). Burnet's translation, on the contrary, is correct: 'And I shall tell thee another thing. There is no substance of any of all the things that perish, *nor any cessation for them of baneful death*. They are only a mingling and interchange of what has been mingled. Substance is but a name given to these things by men' (my emphasis).

32. Ov. *Met.* 15.252–8.

33. Lucr. *DRN* 1.790–7. This resonates *verbatim* with the criticism of Heraclitus in *DRN* 1.670–1, quoted *supra*, note 12. See also *DRN* 2.750–4; 3.519–20. On Lucretius' criticism of Heraclitus, Empedocles and Anaxagoras, Tatum 1984: 178: 'Two criticisms are common to the physics of all three: they deny the existence of the void and *the primary substances which they propose are perishable*' (my emphasis). In the wake of Sedley 1998, Empedocles is increasingly seen as a source for Lucretius; see for instance Nethercut 2017. The intertextual resonances between Lucretius and Empedocles are compatible, however, with their disagreement about matter and, more importantly, nothingness. On such compatibility, see Garani 2007: 19–21. For nuanced discussion of points of contact and distance, Gale 1994: 56–75.

34. Garani 2007: 88.

35. Ov. *Met.* 15.459–62.

36. The Cyclops' cannibalism: Ov. *Met.* 15.90–5.

37. Ov. *Met.* 15.463–7. On the distinction of *sanguis*, blood as a vital substance, flowing inside, or out of, a living and lived body, on the one hand, and *cruor*, the juice of raw flesh, which is shed in violence, wounds and death, on the other, see Mencacci 1986. For a synthetic overview, Dan 2011.

38. Ov. *Met.* 15.476–7.

39. Luciani and Coletti 1981.

40. Cic. *Luc.* 124.

41. Cic. *Tusc.* 1.18; Emp. 105. On the Empedoclean source: Souilhé 1932. Luciani 2018 places the assimilation of the soul to blood at the core of a transcultural 'archaic anthropology'. Greek and Roman thinkers attribute such an idea to Empedocles in the first place.

42. Lucr. *DRN* 13.42–50.

43. Hipp. *Nat. Hom.* 6; Hipp. *Morb.* I.1.30.

44. Pl. *Ph.* 96 b.

45. Ar. *De An.* I, 405 b. See also Philo of Alexandria, *De Somn.* I 30; Lact. *Opif.* 17, 2 ; Macr., *Somn.* I, 14, 19. On these theories, Mansfeld 1987; Congourdeau 2007: 150–62; Duminil 1983.

46. D. L. 8.1.30–1.

47. Pomona's orchard: Ov. *Met.* 14.623–97.

48. Ov. *Met.* 1.102–6. On the connection between the Golden Age and alimentary nonviolence, between Virgil and Ovid, see Habinek 1990.

49. Ov. *Met.* 1.123–4.

50. Ov. *Met.* 15.96–103.

51. Ov. *Met.* 2.340–66. The Heliades are paradigmatic of Ovid's transformations of humans into trees. See Perutelli 1985. On the 'congruence' of women and trees in this exemplary metamorphosis, see Casanova-Robin 2009. On the sensitivity and sensibility of these plants, and more generally on metamorphosis as a 'preservative event', see Zatta 2016. Emily Gowers in this volume draws attention to what distinguishes human and vegetal bodies.

52. Ov. *Met.* 10.298–502.

53. Ov. *Met.* 1.344.

54. Ov. *Met.* 9.345.

55. Ov. *Met.* 15.8.760–7: *cuius ut in trunco fecit manus impia vulnus, haud aliter fluxit discusso cortice sanguis, quam solet, ante aras ingens ubi victim Taurus concidit, abrupta cruor e cervice profundi.*

56. On Ovid's metamorphosed plants versus Virgil's Polydorus, changed in a bleeding and speaking bush of myrtle (Verg. *Aen.* 3.28–9), Smith 1997, Casali 2007.

57. Ov. *Met.* 14.623–97 (Pomona); Ov. *Met.* 8.684–8 (Baucis and Philemon); Ov. *Met.* 1.89–112 (the Golden Age). I have demonstrated the regularity of this pattern, also discussing apparent exceptions, in Sissa 2019.

58. Ov. *Met.* 10.493.

59. Ov. *Met.* 5.425–37.

60. Scheid 2012: 95.

61. Scheid 2011: 95 (my translation).

62. Ov. *Met.* 15.483. On Pythagoras and Numa: Humm 2014.

63. Plut. *Num.* 8.8.

64. Feldherr 1997: 51.

65. Feldherr 1997: 53.

66. Feldherr 1997: 53.

67. On representations of sacrifice in the ancient literary tradition, see Barchiesi, Rüpke and Stephens 2004.

68. Ov. *Met.* 8.684–8. On the alleged Callimachean sources of the aborted sacrifice, Griffin 1991: 63.

69. Ov. *Met.* 3.10–26.
70. Ov. *Met.* 12.24–38.
71. Ov. *Met.* 13.447–87.
72. Segal 2001: 87 claims that it is only Pythagoras who does so: 'Pythagoras builds up an atmosphere of physical horror around eating and calls attention to the bodily grossness of carnivorous practice, as if we, like wild beasts, devour raw limbs and organs, and "enjoy the blood" of torn flesh (*dapibus cum sanguine gaudet*)'.
73. Feeney 2004: 16.
74. Ov. *Met.* 15: 'What an evil habit he is forming, how surely is he impiously preparing to shed human blood, who cuts a calf's throat with the knife and listens all unmoved to its piteous cries! Or who can slay a kid which cries just like a little child, or feed on a bird to which he himself has just given food!.' See D.L. 8.1.36.
75. Green 2008.
76. Green 2008: 47–8.
77. Ov. *Met.* 15.96–103.
78. Ov. *Fast.* 1.334–6.
79. Scheid 1998.
80. Ov. *Fast.* 1.317–457. On Roman sacrifice in general, see Prescendi 2007.
81. Ov. *Met.* 15.60–142.
82. Ov. *Fast.* 1.361–2.
83. Ov. *Fast.* 1.383–4.
84. This corroborates Ovid's critical attitude toward Augustan politics. On this thorny debate: Kennedy 1992, Sharrock 1994; Barchiesi 1994 and 1997; Davis 2006; Casali 2006; Martin 2009; Vial 2013; Claassen 2016; Giusti 2016.
85. Ov. *Fast.* 1.13.

CHAPTER 2
OVID'S GAIA: MEDEA, THE MIDDLE AND THE MUDDLE IN THE *METAMORPHOSES**

Marco Formisano

But metamorphosis operates against all houses.

Emanuele Coccia, *Metamorphoses*

This chapter is devoted to the powerful appearance of Medea in the middle of Ovid's *Metamorphoses*. I will use some strands of environmental criticism in order to uncover the allegorical potential of this episode, not least in virtue of its central position within the *carmen perpetuum*.

In the summer of 2005, I attended an inspiring production of Euripides' *Medea* by the German director Peter Stein at the Greek theatre in Syracuse in Sicily. Located at the centre of the stage was a little wooden house, which quite transparently symbolized the *oikos* as the centre of family life; it remained there until the very end, when, suddenly and quite spectacularly, it exploded with a terrible roar. Here too the symbolism is clear: having killed her own children, Medea represents the explosion of familial and human order. But her extraneousness to human values can also be read, as this chapter will illustrate, as the alienation of Gaia from humanity after the advent of the Anthropocene. The quotation from Emanuele Coccia's *Metamorphoses* in the epigraph nicely connects metamorphosis and the lack of a home, determined by a perpetual motion and migration.[1] As I will suggest, Medea's flight can be read as a detachment from human affairs and from a focus on 'autochthony'[2] and on earthly life in general.

Before turning to the passage in the *Metamorphoses* with which I am concerned it is first appropriate to clarify from which perspectives I intend to explore the particular textuality of the *Metamorphoses* and its strong relationships to ecocritical thought. One of the most interesting consequences of the emergence of the concept of the Anthropocene in cultural studies and in our more general perception of the world is what David Wood, a philosopher of ethics who describes the impossibility of taking a 'step back' from the world around us, calls a 'loss of externality'. While in the Western tradition 'nature' has long been conceived as something external, located outside, today we find ourselves confronted with the impossibility of identifying an outside altogether: 'Now,' Wood writes, 'there is no outside, no space for expansion, no more *terra nullius* [...] no "out" or "away" as when we throw something "out" or "away".'[3] This loss of externality is paramount, because it has serious consequences for humans' approach to their planet. Making sense of things is always, by definition, supported by separating, excluding, or singling things out, so that 'when that externality is no longer available, we are in trouble'.[4] Interconnectedness represents one of the most discussed aspects in ecocriticism. It becomes now evident that texts describe their universe precisely in terms of profound, i.e. not always visible, interconnection between, for instance, human and non-human, or more-than-human, agents.

And it is not only nature that is always sticking with us. Timothy Morton has developed the fascinating concept of 'hyperobjects', i.e. phenomena characterized by their ubiquity, unavoidability and stickiness. Hyperobjects are responsible for the end of the world that, in Morton's view, has already occurred, i.e. the end of the *previous* world. More importantly, acknowledging the presence of hyperobjects implies a completely different perception of the world, since these 'seem to force something on us, something that affects some core ideas of what it means to exist, what Earth is, what society is'.[5]

Hyperobjects, I would add, modify our perception of what texts can be. Both nature and non-human agents are always part of our world, but also of us; it has become evident that *we* are all interconnected. Looking at any text, but especially at Ovid's *Metamorphoses* from this perspective, i.e. considering its textual universe as an immense 'inside' where everything is *somehow*, in a fuzzy and opaque manner, connected with something else, reveals an undeniable truth about the poem which nonetheless disturbs all our usual hermeneutic categories. We, as readers, are in trouble, compelled as we are to face metamorphosis as an ubiquitous and viscous hyperobject. And it is on the basis this trouble and sense of discomfort that I intend to build the discussion that follows.

To be sure, the adoption of an ecocritical perspective for the investigation of ancient texts is by no means unproblematic. It compels interpreters to revise deeply and, perhaps, even abandon many of their usual hermeneutical attitudes. Originally, environmental criticism was conceived with an urgent political agenda: by focusing on the relationship between literature and the environment, this form of critical thought places nature and the earth at the centre of its hermeneutics, in a similar fashion as, for instance, feminism draws attention to gender, or Marxism to capitalism and all its implications.[6]

In the field of classics, texts are very often read as documents and sources, for the reconstruction not only of historical and political aspects, but also of literary history and aesthetics. In other words, it is almost unconceivable for a scholar of Graeco-Roman antiquity to read an ancient text without being particularly attentive to its specific historical and cultural context. Environmental criticism not only offers the opportunity of adopting a different critical language and a different methodology, but more importantly it is able to reveal aspects of the text which usually remain hidden beneath the surface of contextualizing and intertextual readings. In other words, ecocriticism is not just another hermeneutic tool to add to others, but it has the potential of radically changing our perception of texts and literary criticism altogether.

The question whether an environmental awareness existed in the ancient world, or whether ancient authors were interested in consciously thematizing problematic relationships with their environment and more-than-human agents, is certainly legitimate and deserves attention. Here, however, I adopt an ecocritical perspective as a lens through which to look at an ancient text in a radically different way. Reading ancient texts as anticipating current environmental preoccupations is undoubtedly a productive technique, as many scholars have carefully shown.[7] And yet on closer examination epistemological and methodological problems arise when considering this contemporary theoretical approach together with ancient texts. Ancient environmental historian Schliephake expresses worries about how to prevent 'the risk of approaching the distant worlds of antiquity anachronistically, and to impose our own standards and concepts all too freely on societies with different technological, religious, and social backgrounds'.[8] Schliephake, who has made valuable contributions to the study of the environmental humanities and the ancient world, also makes clear that there is a certain similarity between the narrative propagated by scholars of the Anthropocene and ancient and early modern historiographical models of rise-and-fall; thus, he concludes, 'culturally speaking, an Anthropocene existed long before its geological and material effects became apparent'.[9] Nonetheless he is wary of extending this comparison between our current ecological crisis and ancient culture because of the intrinsic danger of anachronism.[10]

Schliephake's approach to the methodological problem is exemplary of classicists' relationship to their texts, which fundamentally aims at contextualizing and reconstructing. But this caution, especially from the perspective of a literary scholar, is problematic for a number of reasons which cannot be discussed in the brief space of this chapter, but one in particular deserves to be briefly mentioned. As has been magnificently shown by Dipesh Chakrabarty, considering the Anthropocene always implies a departure from *human* history: the discourse of Anthropocene compels us to dismantle previous conceptions of history, since the Earth has its own temporality, that is incommensurable with our human history.[11] In other words, if we take seriously the advent of the Anthropocene and carefully consider its consequences, what we tend to conceive as historical accuracy becomes almost an insignificant element, especially from the perspective of literary interpreters. The questions then become, instead, whether we as modern readers, preoccupied with the current environmental crisis, can identify a prefiguration of our situation in ancient and premodern texts. Morton suggests, moreover, that 'we are only just beginning to think the ecological thought';[12] whatever exactly we call it, this thought can become a constant companion with which to approach our world, and our texts. It is a totalizing but also liberating mode: 'the more you consider it, the more our world opens up'.[13] Its power is ubiquitous, since 'it affects all areas of thinking'.[14] Following Morton's path, Virginia Burrus points out that ecological thought, while propelling us 'forward toward a future as yet barely imaginable', also draws us 'back to a past only dimly recalled'. But this past is by no means a 'purer time and place', which would not help us in any case: 'Rather, the pursuit of a usable past here evokes a context as complex and in its own way as compromised as our own.'[15] Burrus devotes her attention to early Christian texts and culture, but the question urges itself on us for every pre-industrial past. How, then, can we *think ecological thought* with and through Ovid? Or how can we *think* Ovid through ecological thought? Thinking the Anthropocene as a moment of profound disjunction between history and future, in the way Chakrabarty has taught us, implies a radical rethinking of literary hermeneutics, especially in critical work on ancient texts.

As has been shown with great clarity by Elizabeth DeLoughrey in *Allegories of the Anthropocene*, it is a task of ecocritics to look for narratives and representations that can help us navigate our ecological crisis. The Anthropocene is not only a material effect, she argues, but also a representational one, since 'it raises vital questions as to how the planet as a system can be signified'.[16] In considering the profoundly disjunctive relation between humans and their planet caused by the rise of the Anthropocene, allegory plays a fundamental role in literary hermeneutics since it is able to catch the enormous difference between the planetary and human scales. The clash between human and planetary scale being a universally valid truth, readers should be able, for instance via allegoresis, to find in every text the incommensurability of the Anthropocene with human history. This disjunction between humans and their planet has been variously discussed and emphasized. The 'intrusion of Gaia', for instance, represents a completely new feature of our own cognitive and hermeneutic horizon. In the words of Deborah Danowski and Eduardo Viveiros de Castro:

> This is the most significant phenomenon of the present century: 'the intrusion of Gaia' (I. Stengers), brusque and abrupt into the horizon of human history; the sensation of a definitive return of a form of transcendence that we believed transcended, and which reappears in more formidable form than ever. The transformation of humans into a geological force, that is, into an 'objective' phenomenon or 'natural' object, is paid back with the intrusion of Gaia in the human world, given the Earth System the menacing form of a historical subject, a political agent, a moral person (Latour). In an ironic and deadly (because recursively contradictory) inversion of the relationship between figure and ground, the *ambiented* becomes the *ambient* (or 'ambienting'), and the converse is equally the case. It is effectively the collapse of an ever more ambiguous environment, of which we can no longer say *where* it is in relation to us, and us to it.[17]

Why should we, readers and interpreters of ancient Greek and Latin literature, not look for Gaia's intrusion within our texts? Some might consider this approach to be an appropriation of ancient cultural products unrelated to current ecological thought. My response would be that not only are many current interpretative models based on *aggressively* modern perspectives (and, as far as I am concerned, rightly so!), although they are presented with a reconstructive rhetoric typical of the discipline, but also that this appropriation is necessary, since it represents a regenerative critical response, political in broader terms, to the perennial denial of Gaia within Western culture.

In what follows I adopt this perspective by departing from the intertextual mode of reading which so heavily characterizes interpretation in Ovidian studies. The *Metamorphoses* is arguably *the* ancient text which more than any other lends itself to ecocritical interpretations – for a number of reasons, but especially because the metamorphic script itself establishes a continuity amongst different natural entities (humans, plants, rocks, earth, flowers, rivers etc.) and also because its textuality, with its rhizomatic multiple structure, precisely parallels the complexity of environmental discourses. This consideration is particularly productive for interpretation of a particular episode of Ovid's poem, with attention to its central position in the seventh of the poem's fifteen books, its protagonist Medea, and its subject: metamorphosis through magic. As has been noted by various interpreters, the appearance and consequent disappearance of

Medea is both astonishing and troubling. For Ovid's Medea does not correspond to Euripides' tragic heroine nor to Apollonius Rhodius' enamoured young girl. Moreover, in the preceding literary tradition she had not been a metamorphic figure, as observed by Ingo Gildenhard and Andrew Zissos in their enlightening study 'The Transformations of Ovid's Medea'.[18] How, then, can this specific heroine contribute to an ecocritical reflection on the *Metamorphoses*? Medea's appearance presents many aspects that can be seen together with the 'intrusion of Gaia' in our world as described by Isabelle Stenger, Bruno Latour, Deborah Danowski and Eduardo Viveiros de Castro among others. Bringing together an ancient mythological and literary figure and a contemporary concept might seem a rather audacious experiment, but this kind of comparison has been persuasively theorized by Donna Haraway, who invites us to engage in a 'diffractive reading'. Similarly to the optical phenomenon of diffraction, wherein – as opposed to reflection – perception is shaped by interference patterns, diffractive reading undoes the historical order and/or aesthetic hierarchy between two or more texts, art works, concepts or events.[19]

Ovidians have sometimes asked the question as to what is at the centre of the *Metamorphoses*. The very middle of the *carmen perpetuum*, represented by Books 7 and 8, seems to elude an explanation of its central position.[20] Adopting a metaphor from Latour, these books can be said to represent the 'muddle' within a poem that readers perceive as classically conceived. Classicists often tend, in fact, to construct their own textual 'providence', i.e. a teleological hermeneutics where everything comes nicely together without contradiction or conflict. But as Latour notes, speaking of Gaia:

> the very idea of providence is blurred, pixellated, and finally fades away. The simple result of such a distribution of final causes is not the emergence of a supreme Final Cause, but a fine *muddle*. This muddle is Gaia.[21]

Along these lines, by considering the poem as a universe in its own terms, an environmental whole or an *Umwelt* in relationship to which no outside is conceivable, I propose to establish the figure of Medea at the significant centre between key moments of Book 1, such as the impossible description of the Chaos, the destruction of humanity through the flood and their rebirth from the earth with Pyrrha and Deucalion, and Book 15, characterized by Pythagoras' appearance. These three moments thematize and represent, though in different ways and according to different modalities (literally, metaphorically, allegorically, or also materialistically, mythologically, philosophically), the 'emergence of Gaia', as Latour calls our current 'metamorphic zone' that is becoming visible once more, 'as if under "nature" the world were reappearing'.[22]

Medea is of course a well known mythic figure and a protagonist of Greek literature in various genres, from Pindar through Euripides to Apollonius. Yet Ovid's poem presents a new and strange Medea who, rather than being in contradiction with all the features attributed to her by the previous literary tradition, combines them in a problematic manner. As pointed out by Newlands, this metamorphic Medea 'is not a coherent, rounded character',[23] since the way in which she is presented simultaneously combines all aspects coming from her literary past and alienates her from all of that: her 'story lacks a tragic dimension and moral complexity', although her figure was previously shaped by the dramatist Euripides and the epic poet Apollonius.[24] Structurally, it is important to emphasize that Medea is given a significant space, half of a book, 'a privilege she shares, not coincidentally, with Pythagoras'.[25] She represents a 'watershed figure in the universal history of transformative change that unfolds in the course of the poem'.[26] Although Medea seems to be a privileged figure in Ovid's work at large (one of the *Heroides* is in her voice, and the poet wrote a tragedy *Medea*, now unfortunately lost), in the *Metamorphoses* a new metamorphic paradigm is introduced through her story. She is the first human agent of transformation, and as such who operates according to a 'procedural' paradigm opposed to the divine mode of metamorphosis: the gods transform their subjects instantly without recurring to herbs, drugs, potions and complicated magic procedures. Medea *intrudes* into the metamorphic narrative in order to change it profoundly: the metamorphoses that follow hers are quite unlike those which precede it. This long episode consists of five sections: a description of the young Medea in love with Jason (7, 1–158), accounts of the rejuvenation of Jason's father Aeson (7, 159–293) as well as of the failed rejuvenation which result in the murder of King Pelias (7, 297–351), her chariot ride to Corinth (7, 351–403), and the final scene in Athens (7, 404–24).

The major characteristic of this tale is represented by magic, which has the simultaneous power of superseding and marginalizing both divine and human activity. More importantly, the Ovidian Medea is presented as an 'interstitial character', because she 'lives in the gaps of her own tradition, gaps that are used to supply metamorphic feats and evocations'.[27] This powerful metaphor can be further explored. The interstitial nature of the heroine perfectly resounds with the central position of her appearance within the poem.

A second, more intriguing association can be made with the emergence of a *tiers paysage*, 'third landscape', defined by landscape architect and theorist Gilles Clément as the sum of residual spaces, including places that have been previously abandoned by humanity: in fact, the main characteristic of the third landscape is that it flourishes precisely because humans are not present and do not influence its shapes.[28] Clément does not define this space metaphorically, since he is interested in identifying third landscapes in contemporary, mostly urban, settings. Yet the definition could be productively used in order not only to identify literary descriptions of places that correspond to those characteristics, but also, and more intriguingly, to describe sections of texts characterized by their interstitial nature either for their position or for their presentation.

The episode of Medea seems to fit in this definition, which would be worth exploring in detail throughout the whole poem. Magic represents a *third way*, as its power is located between gods and humans, and Medea's magic, as will be discussed shortly, is fundamentally based on earth's resources. She can easily be called the 'patron saint of metamorphic narrative',[29] but at the same time her transformative power has nothing in common with previous metamorphoses in the poem. It is as if readers struggle with this figure: they can certainly recognize her, but at the same time she reveals unknown aspects. As pointed out by Alessandro Barchiesi, this Medea is driven by an exercise in positioning herself anew after a long series of personifications and roles,[30] and yet it is as if readers are now able to discover the true Medea, able to justify *ex post* her actions as narrated by others, for instance Euripides and Apollonius Rhodius. Newlands even see Ovid's 'segregation as narrator from Medea',[31] i.e. his lack of empathy with her story. A Medea who is therefore incomprehensible even to her own author, who abandons her to her own story? Can this narrative, then, freed from bulky authorial intention, lead us towards a completely different appreciation of this untraditional Medea? I would suggest that this barbarian heroine is in the *Metamorphoses* a figure for a dispossession which is both ethical and methodological. On an ethical level her actions do not follow human and divine norms, and on a hermeneutical level she does not belong even to her own literary past: she is a 'barbarian' to herself. Like readers of this particular episode of the *Metamorphoses*, humans may believe they know their planet – but now Gaia, as Latour suggests, is showing her true face, at the same time familiar and strange, with incomprehensible and terrifying aspects.

The opening lines of Book 7 are highly significant since they bear an important trace of the discourse of nature and earth, which in its materiality is regularly obscured and neglected in scholarly readings of this passage.

iamque fretum Minyae Pagasaea puppe secabant,
perpetuaque trahens inopem sub nocte senectam
Phineus visus erat, iuvenesque Aquilone creati
virgineas volucres miseri senis ore fugarant,
multaque perpessi ***claro*** *sub Iasone tandem*
contigerant rapidas ***limosi*** *Phasidos undas.*

Met. 7.1–6

And now the Argonauts were cleaving the sea with their Pagasaean vessel. They had seen Phineus drawing out his old age in eternal darkness, and the youthful sons of the North Wind had driven off the maiden-bird Harpies from the poor old man's face. Finally, after enduring much under the leadership of the illustrious Jason, they had touched the rapid waters of the muddy Phasis.

Scholarly interpretations of this passage have typically pointed to a range of intra- and intertextual relationships. Gildenhard and Zissos see here a denial of the Argonauts' epic adventure as narrated in the first two books of the Apollonius' poem, and they also notice a tension in poetological terms: 'A central programmatic preoccupation of the *Metamorphoses*, signaled in the exordium, is the aesthetic tension arising from the conflicting demands of a *carmen perpetuum* (referring to large-scale epic composition) and a *carmen deductum* (referring to smaller-scale poetry characterized by Alexandrian refinement).'[32] Barchiesi comments on *perpetua sub nocte* (7, 2) as a programmatic echo of the prologue's *carmen perpetuum*, and although he points out a significant detail, the muddy waterway (7, 6: *limosi Phasidos*), a feature which had never been attributed to this specific river before, nonetheless he considers it 'suspicious': in its apparent lack of connection, he sees in it a *transparent* reference to Callimachean poetological agenda.[33]

Without denying what Ovid himself might or might not have thought, I would note that the claim that the Roman poet wanted to allude to Callimachus or, more specifically, playfully to rewrite Apollonius' poem on the Argonauts under a Callimachean aegis, focuses attention on the author and his intentions rather than on the text itself. To put it more bluntly, from the Latinist's perspective criticism needs to make (supposed) authorial intentions fit with their texts, almost as if the poem were merely a piece of evidence whose purpose was to clarify, assert or even confirm what they think Ovid might have thought. Arguably, this is a delightful intellectual game. Nonetheless, any text is always much more complex than a reduction to the identification of a web of intertextual relationships. In this case, for instance, a simple but revelatory detail indicates much more than what has been established by the acute eye of intertextual critics: the river Phasis is defined as *limosus* ('muddy', 'opaque') while Jason is *clarus* ('illustrious' but also 'bright' and 'transparent', i.e. the opposite of *limosus*), a *muddy* natural element and a *clear* man. Édouard Glissant discusses in his foundational *Poetics of Relation* the opposition between opacity and transparency in literary language. 'Literary textual practice', he writes, 'represents an opposition between two opacities: the irreducible opacity of the text, even when it is a matter of the most harmless sonnet, and the always evolving opacity of the author or a reader.'[34] This tension between transparency and opacity can generate a productive approach to Ovid's text, because opacity 'is not enclosure within an impenetrable autarchy but subsistence within an irreducible singularity'.[35] Seen from this perspective, the 'muddy Phasis' leads the reader to knowledge of Medea not as a stereotypical figure of myth and previous texts, but as an 'irreducible singularity' on her own. Glissant sees in the insistence on transparency a typical trait of Western culture and its inherent imperialistic drive; opacity therefore has the task of unveiling that violence.

> This same transparency, in Western History, predicts that a common truth of Mankind exists and maintains that what approaches it most closely is action that projects, whereby the world is realized at the same time that it is caught in the act of its foundation. Against this reductive transparency, a force of opacity is at work. No longer the opacity that enveloped and reactivated the mystery of filiation but another, considerate of all the threatened and delicious things joining one another (without conjoining, that is, without merging) in the expanse of Relation. Thus, that which protects the Diverse we call opacity.[36]

Moreover, in the exordium of Book 7 territoriality itself is emphasized. Environmental hermeneutics teaches us how to appreciate descriptions of places in their very materiality. The initial reference to the muddiness of the river Phasis, beyond any intertextual reference and discourse on literary genres, should be taken for what it is: the reader is entering the muddy territory of this particular story, located at the significative centre of the metamorphic universe. We can read it as a reference to the muddy earth and its epistemic opacity, opposed to human or, more specifically, the 'clear' or 'bright' reason of a man (*claro sub Iasone*) who wants to determine and subjugate it. Jason, to use Glissant's terms, is configured as the hero of transparency, while Medea is a figure of opacity in many regards: her appearance disturbs the logic of filiation and also, as we will see, of maternity. Her encounter with Jason is highly symbolic in other terms as well. Jason comes from the *sea* and enters the muddy Phasis, i.e. where water and land melt, in order to meet Medea on *land*, but she will then *fly* away.

From a geo-critical perspective, this central episode of the *Metamorphoses* thus materializes what we can call an *earthly epitome*, consisting of water, land and air. 'Land' is definitely what characterizes Medea's configuration within this episode of the *Metamorphoses*. If we apply to Medea the terms of Danowski and Viveiros de Castro mentioned above, we can describe her as passing from an *ambiented* to an *ambient* figure. Or, as Aït-Touati and Coccia suggest, Gaia is an agent without environment, precisely because she *is* her own environment.[37] For through this episode the readers witness the collapse of the distinction between the cosmologic and the anthropologic order. Magic, i.e. Medea's power, is a marker of this collapse. Danowski and Viveiros de Castro describe the Anthropocene as 'the age when geology has come into a properly geological resonance with morality':[38] of this process of 'geologization of morals' Medea stands as an exemplary literary embodiment. Her initial opaque monologue brings to the stage several elements that can be read in light of the conceptual association between the heroine and the land that is variously thematized throughout the entire episode. Right at the beginning, for instance, when she first meets Jason and instantly falls in love with him, she warns herself: *haec quoque terra potest, quod ames, dare* (7, 23) ('this land too can give you something to love'), where land and love are put in conceptual continuity. Later, when Medea is still uncertain whether to follow Jason, she says:

> *ergo ego germanam fratremque patremque deosque et*
> *natale solum ventis ablata relinquam?*
> *nempe pater saevus, nempe est mea barbara tellus*
>
> *Met.* 7. 51–3

> My sister, brother, father, gods, native soil – shall I abandon them, carried off by the winds? To be sure, my father is cruel; to be sure, my land is barbarian.

Medea's *natale solum* (her 'native land') becomes *barbara tellus*. The land is at one and the same time theatre of the action and what is at stake. Jason has to fight against enemies and bulls born from the earth (*tellure creatis / hostibus*, 7, 30–1 and *cur non tauros exhortor in illum terrigenas feros*, 7, 35–6) so that he can eventually go to sea again (*det lintea ventis*, 7, 40). Land and its tension with both sea and air remain central to the entire episode.

As Gildenhard and Zissos observe, in this Ovidian version of the myth 'Medea fuses her epic present with her tragic future' by anticipating 'arguments that Euripides' Jason uses at a chronologically subsequent stage'.[39] This is certainly true; but there is more to say. By calling her own land 'barbaric', for example, she demonstrates a global awareness, as she is able to look upon her own *Umwelt* from the perspective of an outsider. For her land is barbaric only if seen from Greece, a place which at this point in her story she does not yet know. As we have seen, according to Wood our current global situation is marked precisely by a 'loss of externality'; Medea seems to incorporate and include the external in her vision. In doing so, she alienates herself from her family (*germanam fratremque patremque*) and her own territory, depriving it at the same time of her own ethical commitment and humanity (*barbara*). This sense of alienation from herself characterizes her speech from the beginning. She seems to take distance even from language when she says: *mirumque nisi hoc est / aut aliquid certe simile huic, quod amare vocatur*: 'I wonder whether this, or something like it, is what is called "loving"' (7, 12–13). Not only does she allude to the fact that her feeling for Jason might be 'love', a feeling she has never felt before, but she describes the process (*amare*) rather than the result (*amor*, love itself). In this sense, she becomes a stranger in her own land and language: alienation and defamiliarization, as discussed, for instance, by Niccolò Scaffai, represent some of the most frequently occurring literary devices when the difference of scale within the environment is emphasized in texts.[40] In this case, however, it is Medea herself who strangely alienates her own present from an environment and context which, at this point, she can only take for granted. She seems to embody Glissant's 'poetics of relation', since she is able not only to acknowledge the other in its diversity and intangibility, but even to assume the other's position. Glissant makes an interesting distinction: 'the verb *to grasp* (*comprendre*) contains the movement of the hands that grab their surroundings and bring them back to themselves. A gesture of enclosure if not appropriation.' To 'grasping' he prefers 'giving-on-and-with' (*donner avec* in French) that 'opens on totality'.[41] Eventually what Medea decides at the end of her long monologue devoted to understanding her feelings and her situation – in order to avoid committing a crime (*effuge crimen*, 7, 71) – will be completely subverted by subsequent actions.

Land continues to be paramount in the following scenes. For the properties of earthly substances are a central feature of the techniques of rejuvenation practiced by Medea, (7, 179–296): they both represent the metamorphic virtue of the heroine and thematize the power of earth as such. Central to the figure of Medea in Ovid is the fact that she possesses a knowledge of transformation, and this knowledge is very much based on complex procedures. As has been acutely observed by Gildenhard and Zissos, the potency of the herbs which Medea uses 'suffices to trigger a "natural" metamorphosis', and yet they interpret the careful description of the procedure and enumeration of various ingredients from the perspective of the discourse of genre, i.e. as 'a calculated breach of epic poetry'.[42] But it is also important to emphasize the numerous references to earth, both direct and indirect, and to animals. The earth is the gravitational centre of the entire magical operation, and the moon opens the scene by looking at the earth (*et solida terras spectavit imagine luna*, 7, 181). During the night, when all things and all animals freeze in a silent and motionless atmosphere, Medea goes 'barefoot' (*nuda pedem*, 7, 183) and invokes the moon by 'kneeling on the hard earth' (*in dura submisso poplite terra*, 7, 191). The earth is then directly addressed as the one who teaches the magicians how to use the herbs (*quaeque magos, Tellus, pollentibus instruis herbis*, 7, 196). The subsequent description and enumeration of herbs and animal ingredients is full of references to the earth (e.g. *de caespite* 240; *tellure* 241; *fossas* 245; *terrena numina* 248; *stratis in herbis* 254; *radices* 264; *seminaque floresque et sucos incoquit atros* 265; *harenas* 267; *in terram* 283; *humus* and *mollia pabula* 284). Almost at the end of this long passage, describing at length Medea's complex procedural magic, she adds 'a thousand other nameless things' (*sine nomine rebus*, 275). Newlands interprets this detail as an intentional touch of humour on Ovid's part who at this point, seemingly exhausted after the long description, takes a certain distance from his own heroine's magic,[43] while Gildenhard and Zissos speak of 'inexpressibility', emphasizing that those nameless items 'are beyond the reach of language' and therefore available only to Medea's exceptional magic knowledge.[44] The passage is rich in details:

> *his et mille aliis postquam sine nomine rebus*
> *propositum instruxit mortali barbara maius,*
> *arenti ramo iam pridem mitis olivae*
> *omnia confudit summisque inmiscuit ima*
>
> *Met.* 7. 275–8

> After the barbarian woman crafted her plan, something more than mortal, with these and a thousand other nameless things, she stirred it all up using a branch of kindly olive long parched, mixing the bottom with the top.

Not only the expression *sine nomine* but also the generic *res* is revealing; the latter points to a non-human earthly universe where (human) language is banished because here its basic function of naming and describing *things* can have no place. *Res* refers not only or necessarily to herbs but leads the reader into an earthly universe which remains indistinct, unknown and opaque for humans and, more importantly, powerfully exists quite independently of them. Medea is accordingly called 'the barbarian woman' (*barbara*): alien, and as such able to operate on a level outside of the human scale, not through some divine qualities but precisely through her supreme mastery and sovereign control over the earth and its mysteries. She is eventually able to bring about what can allegorically be read as a powerful cosmic revolution: she mixes everything up, *summa* with *ima*, heavenly with earthly *things* (7, 278). The progressive 'geologization' (Danowski and Viveiros de Castro) of Medea as a cosmic force within the metamorphic universe is also established by another macroscopic aspect which has been identified by Gildenhard and Zissos: her story decisively 'foregrounds the topographical at the expense of the chronological'.[45] In other words, time is absolutely secondary, while space becomes the cypher of Medea's incursion in the metamorphic *Umwelt*. As Latour comments, while traditional conceptions see nature as consisting of strata and levels which it is possible to go through in an orderly manner, Gaia now subverts all levels (7, 278 *omnia confudit summisque inmiscuit ima*): 'There is nothing inert, nothing benevolent, nothing external in Gaia,' and, since climate and human life now become tightly connected as never before, 'space is not a frame, not even a context: *space is the offspring of time*'.[46]

If, in virtue of a diffractive reading as described by Haraway, we consider the Ovidian Medea in the light of Latour's Gaia, it is possible to uncover the potency of Medea, significantly located at the core of the *Metamorphoses* as 'a figure of fundamental importance in Ovid's cosmic history',[47] and to reconsider her relationship to humans. It is Gildenhard's and Zissos' contention that Medea in this poem represents 'by far the most prominent *human* agent of transformation'[48] because of her magical skills and knowledge. Yet precisely her ability to appropriate secret and otherwise nameless earthly elements (7, 275, *sine nomine res*) makes her radically different from humans; she is arguably neither a goddess nor a human. In virtue of both her 'interstitial nature' and her magic, she represents a third way, such that she can actually do something that even gods cannot: to extend human life. In this regard, an often ignored detail is particularly revealing. After Aeson's rejuvenation, Bacchus is said to have observed and admired this procedure and to be eager to apply it to his old nurses: *viderat ex alto tanti miracula monstri / Liber* (7, 294–5). Bacchus watches 'from above' the earthly *monstrum*, a rather frightening and unheard-of miracle which is evidently otherwise unknown to gods as well as mortals.[49]

Medea locates herself between humans and gods within an earthly environment, and she works with things that apparently escape both divine and human language (*sine nomine res*, *monstrum*). As such, in Glissant's terms, she is a figure of both opacity and relation in which everything converges but at the same time irreparably differs. Also, this Medea seems to be deprived precisely of a characteristic for which her personage was famous in the preceding literary tradition: her actions systematically escape psychologization. As has been pointed out, Ovid's readers remain extraneous to the motivations of her enterprises. For example, soon after the successful rejuvenation of Aeson, we read that she duplicates the procedure but in this case in order to kill king Pelias. The motivation provided lacks any rational explanation: *neve doli cessent* (7, 297, 'so that her acts of deceit might not cease'). Medea seems to be deprived of a strategy, pushed simply by a murderous drive. The same is true later, when her infamous infanticide is only briefly alluded to in two verses (7, 396–7): the Ovidian text does not provide us with any sort of reference to the powerful Euripidean antecedent and her exemplarily nuanced internal drama. We as readers concentrate 'more on what Medea does, and less on why she does it'.[50] She operates 'in an ethical vacuum',[51] within which there is no room left for any form of introspection or moral empathy.

This aspect emerges particularly in the second section of the episode, when Medea, after having caused Pelias to be killed by his own daughters (who are deceived by her rejuvenation procedure), goes on a tour which will lead her in the end to an airborne exit (7, 350–424). The route of Medea's voyage parallels that of the Argonauts,[52] but as Gildenhard and Zissos note, her flight over the Aegean is deprived of any reason or rationale, nor does it seem to be driven by a specific intentionality: 'Medea is a sightseer, a tourist, who remains aloft for the duration and performs no concrete task.'[53] Moreover, her mastery of space is underscored by the fact that the various metamorphoses which are narrated at different stations during her trip do not have any particular significance, whether for her or for the narrative structure of the episode or the poem as a whole: they seem only to emphasize the local landscape. Gildenhard and Zissos here emphasize metaliterary and intertextual elements; observing the lack of a causal or temporal structure within this sequence of unconnected episodes, they identify a specific strategy chosen by the author, consisting in presenting various metamorphoses as a sort of Hellenistic catalogue, in Callimachean style. This sequence of disconnected episodes is the opposite of Ovid's *carmen perpetuum*, designed to offer a continuum of interconnected stories; in this sense, they observe, Medea's catalogue of vignettes 'emerges as a veritable anti-*Metamorphoses*'.[54]

But beyond any intentional design on Ovid's part, the text also suggests that Medea during her aerial journey produces another narrative, one which consistently lacks temporal structures. To be more precise, not only is there properly no narrative (if narration is understood as a sequence of logically connected scenes) but also no particular authorial agency. Medea is a simple tourist, and as such she is not in charge and does not determine what she observes from her chariot while riding through the sky. Her function seems precisely to be the undoing of human history and consequently literature, not only in terms of a specific narrative strategy taken by Ovid himself. During her flight, Medea appears extraneous, not interested or concerned in the least with what she is seeing. But above all it is motion itself that is thematized. The lack of narrative coherence and the difficulty of finding any apparent meaning within the episode suggest that Medea's flight deserves our attention on its own and as such. Perpetual movement, as suggested by Coccia, characterizes the life of and on our planet, a word that derives from Greek *planaomai*, meaning 'to wander': 'It is because of the *planetary* nature of Gaia and all of her children that every body on Earth is subject to metamorphosis.'[55]

Among the various metamorphic episodes narrated during Medea's flight, one of the first is striking for the reference to the flood narrated in Book 1.

. . . fugit alta superque
Pelion umbrosum, Philyreia tecta, superque
Othryn et eventu veteris loca nota Cerambi:
hic ope nympharum sublatus in aera pennis
cum gravis infuso tellus foret obruta ponto
Deucalioneas effugit inobrutus undas.

Met. 7.351–6

She fled on high, over shady Mount Pelion, home of the centaur Chiron, and over Mount Othrys, the region made famous for what happened to old Cerambus: assisted by nymphs, carried up into the sky on wings when the heavy earth had been deluged by the sea, he escaped Deucalion's waters undeluged.

This short reference to the genesis of the human race after the apocalyptic flood has confounded critics. Not only does it not have any logical connection to Medea's story itself, being presented merely as a gratuitous detail; but also, and more significantly, Cerambus' *eventus* openly contradicts the initial narrative of Pyrrha and Deucalion in Book 1, where Ovid explicitly tells us that the couple are the only human beings who survived the flood sent by Jupiter in order to destroy humanity (*Met.* 1.325–6). This offering of two or more versions of a myth is a recurrent element in the *Metamorphoses*, but here this cannot be reduced to a purely narrative game. In this case, despite the lack of a logical connection, it is easy to observe a similarity between Cerambus and Medea: both are saved by an aerial intervention so that they can fly safely away from the place where their life is at risk. This similarity establishes a sort of allegorical connection between the two figures right at the beginning of Medea's chariot-ride, as if Medea, while watching from above, were able to recognize a part of her own destiny. But the allegorical relationship is also subverted: here it is the allegorized mythical subject, i.e. Medea, who looks down to the allegorizing object, the human Cerambus. More importantly, the contradiction with the 'official' narrative established in Book 1 has the effect of arousing fundamental doubts about the veracity not only of Ovid's tale of anthropogenesis (which is directly evoked by *Deucalioneas*),[56] but more generally by destabilizing what within the universe of the poem are described as the origins of human history. And it also brackets this crucial cosmogonic event by rendering it as only an insignificant detail, offered to readers *en passant* and emphasizing their fundamental relativity, and perhaps even irrelevance, from the perspective of Medea, who observes without commenting.

At closer glance there is another interesting detail. This surprising new version of the deluge openly challenges not only Ovid's account but also what Jupiter himself is said to have done. It is, however, not Medea herself who produces this new version: she simply observes it. Therefore, it is as if she were testifying to another truth, one which does not correspond to the one authorized by the poet in Book 1. But while Ovid's account is presented as authoritative, Medea's is a vision from above, emotionally detached and uninterested. Her aerial journey undoes not only the internal Ovidian narrative but also human temporality, and represents pure movement in space bereft of purpose: 'Medea's narrative has no teleology as such.'[57] This lack of human teleology is precisely the main characteristic of the Anthropocene, which introduces another kind of knowledge, or better, a different approach to knowledge, one based on geology and natural time rather than on human history and ethics. The Anthropocene and the 'intrusion of Gaia' cause a radical change in the perception of the world, a fundamental incommensurability of the environment with the human vision of it. Medea's aerial perspective represents this incommensurability between human and planetary scale: her attention from above is still devoted to events happening down on the earth.

This profound disjunction seems to characterize Medea's actions. The lack of any teleological intent, not measurable on a human scale, is also represented by Medea's inconclusive exit: she simply disappears, leaving the text, and her own story, through the power of *carmina*, i.e. on her own terms, mastering even climatic phenomena (7, 424, *effugit illa necem nebulis per carmina motis*, 'she escapes slaughter, stirring up mists with her spells'). As observed by DeLoughrey, the monumental scale of the planet always challenges our own perception, which ends up being inadequate. In order to trace this inadequacy in literary representation we must then activate allegory and allegoresis, that is the hermeneutic possibility of making sense of the disjunction between what is said and what is meant.[58]

The miniaturization of important or well-known moments is a characteristic of the entire passage in Ovid's poem. More generally, dimensions are a significant feature of the narrative, pointing to uncanny and unexpected aspects of stories that are otherwise told and retold. The most representative passage in this regard is the colossal *mise entre parenthèses* of Euripides' plot, and of the famous infanticide in particular, in only four lines:

> *sed postquam Colchis arsit nova nupta venenis*
> *flagrantemque domum regis mare vidit utrumque,*
> *sanguine natorum perfunditur inpius ensis,*
> *ultaque se male mater Iasonis effugit arma.*
>
> *Met.* 7.1–6

> After the new bride had burned with Colchian poisons, and each of the two seas at Corinth had seen the royal palace in flames, a wicked sword is drenched in children's blood, and the mother, ill avenged, escapes Jason's arms.

Postquam: '*after* the new bride burned', Ovid refers to Medea's infanticide *en passant* and with a passive verb (*perfunditur*) without agent, as if it were not only a universally known fact which needs no elaboration here, but also an event which is, after all, unnecessary for the plot, and which is some way explicable in view of her previous story told in the first section of Book 7. But if it is true that Medea's famous role as infanticide is minimized and reduced to a mere detail in a long series of crimes and undertakings, it is also a fact that precisely her role as mother is problematized, to say the least. To be sure, she is a mother in these lines (as the threefold assonance of *ma* in *male mater . . . arma* reminds us), but almost incidentally so: here, her being a mother is not a central feature of her identity.

This passage has generally been read in terms of a conscious allusion to, and reduction or belittlement of, the Euripidean play, but one could go even further: it is the very intertextual nature of this Medea which is downplayed. She may well be *also* the Medea of Apollonius and Euripides, but she is certainly not *only* that Medea. The genealogical structure or cognation between texts as set up by an intertextual reading turns out itself to be minimized here, just as Medea is minimized as a mother. Medea emerges as a powerful force of nature in view of her abilities as sorceress, but also as a denaturalized mother – not so much because she kills her own children, but because her very 'motherness' is questioned by its juxtaposition with *male*.[59] As Sarah Iles Johnston has argued, the figure of Medea in ancient literature is characterized by contradictory traits which bring out problems of self and other, so that 'she became the paradigmatic outsider'.[60] This outsider, stranger to others but also to her own family, shares no common values with anyone. As observed by Newlands, the recurring references to her homeland (*Colchis*, *barbara*) profile her as 'the foreign and outlandish' and 'distance her from common human experience'.[61] Medea emerges here as a figure that disturbs filiation not only because of her paradigmatic infanticide but, more compellingly, because she is barely a mother.

*

I have approached the appearance of Medea in Ovid's metamorphic world as the manifestation or 'intrusion' of Gaia within a human universe. On this reading, Medea becomes a figure which disrupts not only human relations but also intertextuality, the Latinist's darling, a hermeneutical practice directly produced by and expressing the logic of filiation, defined by Glissant as the cultural practice which confirms the power of generalization and the legitimization of the 'single root' (*racine unique*).[62] Medea makes her appearance at the center of Ovid's metamorphic universe as a muddy and opaque figure, a rebel against her own literary past and the logic of filiation, not reducible to an intertextual reading. Transfiguring the concepts of 'home' and 'belonging', she becomes her own environment: from ambiented figure to ambient herself. And, like Gaia, she unveils the 'muddy muddle' of the text – and of our planet, our own *barbara tellus*.

Notes

* Many thanks to the editors of this volume, the anonymous referees, and Simona Martorana for their thoughtful comments; to an inspiring audience at UFRJ in Rio de Janeiro, and to Craig Williams for his help with the translations from the Latin.

1. Coccia 2020: 140.
2. Coccia 2020: 142.
3. Wood 2005: 172–3.
4. Ibid.
5. Morton 2013: 5.
6. See Glotfelty and Fromm 1996: xix.
7. For environmental criticism and classical antiquity see Schliephake 2016 and 2020; on early Christianity Burrus 2019; on early modern periods McColley 2007 and Watson 2006.
8. Schliephake 2016: 3. See Schliephake 2020: 6 for a similar remark.
9. Schliephake 2020: 5.
10. Schliephake 2020: 6.
11. Chakrabarty 2009: 197–8.
12. Morton 2010: 134.
13. Morton 2010: 1.
14. Morton 2010: 4.
15. Burrus 2019: 1.
16. DeLoughrey 2019: 3.
17. Danowski and Castro 2017: 14
18. Gildenhard and Zissos 2013 and Newlands 1997, are fundamental contributions to the analysis of Medea in the *Metamorphoses*. I take these studies as exemplary both for their careful and astute close readings but also for the kind of approach (intertextual and intentional) which I would like to challenge in these pages.
19. Haraway 1997, 273: 'Diffraction patterns record the history of interaction, interference, reinforcement, difference. Diffraction is about heterogeneous history, not about originals.

Unlike reflections, diffractions do not displace the same elsewhere, in more or less distorted form, thereby giving rise to industries of metaphysics. Rather, diffraction can be a metaphor for another kind of critical consciousness at the end of this rather painful Christian millennium, one committed to making a difference and not to representing the Sacred Image of the Same.' See also Gragnolati and Southerden 2020: 3 and Formisano 2021.

20. See Kenney 2011, xi with bibliography. Specifically on Book 8 see Tsitsiou-Chelidoni 2003.
21. Latour 2013: 100.
22. Latour 2013: 63.
23. Newlands 1997: 178.
24. Newlands 1997: 189. Note that Newlands attributes the problematic status of this Medea to Ovid himself, who 'suggests the difficulties and inconsistencies involved in the rewriting of tradition' (p. 191).
25. Gildenhard and Zissos 2013: 89.
26. Gildenhard and Zissos 2013: 89.
27. Gildenhard and Zissos 2013: 90.
28. Clément 2004.
29. Gildenhard and Zissos 2013: 110.
30. Barchiesis 2001: 160.
31. Newlands 1997: 187.
32. Gildenhard and Zissos 2013: 92.
33. Barchiesi 2001: 160.
34. Glissant 1997: 115.
35. Glissant 1997: 190.
36. Glissant 1997: 62.
37. Aït-Touait and Coccia 2021: 7.
38. Danowski and Castro 2016: 15.
39. Gildenhard and Zissos 2013: 125 n. 34. The reference is to Euripides, *Medea* 537: βαρβάρου χθονὸς.
40. Scaffai 2017: 28 uses the term *straniamento* to refer to the 'ribaltamento di prospettive che si attua tra umano e non umano, tra domestico e estraneo'.
41. Glissant 1997: 192.
42. Gildenhard and Zissos 2013: 105.
43. Newlands 1997: 187.
44. Gildenhard and Zissos 2013: 106.
45. Gildenhard and Zissos 2013: 91.
46. Latour 2017: 106.
47. Gildenhard and Zissos 2013: 89.
48. Gildenhard and Zissos 2013: 89.
49. Kenney 2011 in his commentary simply notes that this obscure detail of the myth is added by the poet in order to show his complete mastery of mythological material. This brief comment illustrates the persistent strength of Latinists' focus on authorial intentions, so strong that it sometimes enables them to avoid actually interpreting the text.

50. Gildenhard and Zissos 2013: 97.
51. Gildenhard and Zissos 2013: 109.
52. Newlands 1997: 191.
53. Gildenhard and Zissos 2013: 113.
54. Gildenhard and Zissos 2013: 117.
55. Coccia 2020: 116.
56. As for instance emphasized by Gildenhard and Zissos 2013: 119.
57. Gildenhard and Zissos 2013: 122.
58. DeLoughrey 2019: 15.
59. As Kenney 2011 notes, *male* can be understood both with *ulta* ('ill avenged') and with *mater* ('horrible mother').
60. Iles Johnston 1997: 8.
61. Newlands 1997: 189.
62. Glissant 1997, in particular the chapter 'Expanse and Filiation' (47–62).

CROSS-SPECIES ENCOUNTERS

CHAPTER 3
ANIMAL LISTENING

Shane Butler

A nightingale sings

This chapter[1] offers a close reading of a mysterious Latin poem that, while not by Ovid, is one that belongs not only to his tradition but also to what, in more senses than one, could be said to be an Ovidian 'environment'. The poem is included in the *Anthologia Latina*, a modern concatenation of poems of widely varied dates that are transmitted by various sources. Generally known as the *Elegia de Philomela*, 'Elegy on the Nightingale', the poem is preserved in at least eight eleventh-century manuscripts and many more recent ones. Its ascription to a certain Albus Ovidius Iuventinus by a famously unreliable seventeenth-century scholar and editor, Melchior Goldast, ostensibly on the basis of manuscript collations reported, second-hand, by a sixteenth-century encyclopedia, probably is pure invention (Goldast 1610: 24). (With more forgivable daring, Goldast attempts to geolocate the author to Lombardy on the basis of the poem's use of an unusual word for 'thrush'.) In any case, manuscript evidence for this name is now lacking. A few late manuscripts give the author simply as Ovidius, which probably is nothing more than an attempt to use the anonymous poem to ventriloquize a giant of ancient literature who had also sung of birds. This may in turn have provided the kernel of the correction encountered or concocted by Goldast, once it became clear that the poem could not be by the one and only Ovid of the *Metamorphoses*. For convenience, I shall refer to the author as Albus Ovidius.

There are two redactions of the poem, one longer and one shorter, and a number of textual variants. The poem has been freshly edited (in the longer version) and supplied with a prose translation in a new collection of pseudo-Ovidian poetry for the Dumbarton Oaks Medieval Library (Hexter et al. 2020: 50–5), to which the reader who wishes to consult the entire poem is referred.[2] The poem's seventy-two lines can be divided into four sections: an opening address to the nightingale (lines 1–8), (2) a catalogue of other birds and bird-sounds (lines 9–44), (3) a catalogue of still other animals and their sounds (lines 45–64), and (4) a brief conclusion that first pictures all animals, even mute ones, offering 'gifts of praise to the Lord [*dominus*]' and then ends in apparent address to the poem itself (lines 65–72), which, the author somewhat cryptically hopes, will achieve through transcription (on parchment) an immortality analogous to that of the phoenix. The presumably Christian meaning of *dominus*, along with the reference to parchment, would tend to date the poem no earlier than Late Antiquity, though properly speaking the final lines could have been tacked conveniently onto the end by a Christian reader at any point after the poem began to circulate. (Indeed, the poem's rather abrupt and awkward abandonment of its avian focus after section 2 could be used to argue that sections 3 and 4 are both later additions.) An intriguing if tendentious attempt was made some time ago to link the poem to the times of the Roman co-emperor Geta (i.e., to the end of the first, beginning of the second century CE), who, according to his biographer in the *Historia Augusta* (Geta 5.4–5), used to grill grammarians about the proper Latin words for the sounds made by particular animals, such as *balant* for lambs, *grunniunt* for pigs, and so on (Bernhardy 1850: 294). Others have suggested that the poem, whatever its date, must depend on a scholarly work of the classical period, possibly the encyclopedic *Pratum* (*Meadow*) of Suetonius, of which we possess only fragments (Peck 1894: 228, who provides a helpful glossary of 'onomatopoetic words' in Latin, 230–9). Animal sounds, however, were something of a *topos* in ancient discussions of the origins and nature of language. As we shall see below, they receive especially close attention in the influential treatises of the late antique teachers and linguists known as 'grammarians', a fact which might point to a date in Late Antiquity or beyond; a link has been suggested to a specific but little-known fourth-century grammarian and poet with an attested special interest in animal sounds (Lemaire 1824: 316–17). The most detailed study thus far, however, uses metrical peculiarities to date the poem much later, probably to the tenth century, i.e., not long before its earliest appearance in surviving manuscripts (Klopsch 1973: 174–5).

Strictly speaking, the arguments of this paper do not require settling questions of the poem's authorship, sources, or transmission. Indeed, there is something suggestive about the way that this insistently sonorous text refuses to situate itself securely in literary (or any other) history. (Latin itself, one might say, similarly disrupts neat chronologies by roaming through more than two millennia of surviving texts.) Accordingly, I would like to seize on the poem's various indeterminacies as an opportunity to shift our focus from names and dates to hazier questions of poetry as mediation. By 'mediation' I am not primarily referring to language's ability to *express* this or that signified 'content'. I do partly mean language's ability instead to be *impressed*, that is to say, to be modified in ways that record aspects of its use and context, modifications that are the proper objects of historical linguistics. What, however, I primarily mean by 'medium' is language's ability to *mediate* between human and nonhuman sounds: 'communication', in other words, as a sonic borderland, and even as common ground, somewhere in the middle and in our midst. Albus Ovidius pointedly places his poem in such a space, aiming to make it resonate with and in the human and nonhuman phonospheres that there seem to overlap. 'It is language which speaks, not the author', Barthes famously observes in 'The Death of the Author' (1967), a work that mostly has been received as offering lessons about interpretation, that is, about how a text means. But in context Barthes has just invoked Mallarmé, poet of ceaseless soundplay, who is forever turning us back to the 'genotext' of Barthes's student Kristeva, that is, to the sound-stuff of language, where significance is only ever emergent and unstable (Kristeva 1984: 86). Albus Ovidius takes this one step further. It is not the author who speaks, but neither is it language. Rather, it is the world – a world, *mirabile dictu*, that seems to speak the author's own tongue. In other words, the more successfully the poet mimes the world, the more it seems that it is instead the world that is ventriloquizing both him and his Latin.

Here are the poem's opening lines, with a translation that aims to recall but not to reproduce the metre of the original:

> Dulcis amica, veni noctis solatia praestans;
>   inter aves etenim nulla tui similis.
> Tu, philomela, potes vocum discrimina mille,
>   mille vales varios rite referre modos;
> nam quamvis aliae volucres modulamina temptent,   5
>   nulla potest modulos aequiperare tuos.
> Insuper est avium spatiis garrire diurnis:
>   tu cantare simul nocte dieque soles.

> Come, my sweet friend, bringing night's consolations;
>   no other bird could be quite like you.
> Philomel, voices, all different – a thousand! –
>   one thousand tunes you repeat without fault.
> Yes, other birds make their efforts at music;   5
>   none can come close to your melodies.
> Most birds, besides, prattle only in daytime;
>   you, day and night, regale us with song.

In the genuine Ovid's memorable version of the ancient myth, the nightingale gets its name from a human princess, Philomela, who was transformed into the first of that species, after her brutal rape and ensuing revenge. The horrific violence done to Philomela, which in Ovid includes the removal of her tongue by her rapist, her brother-in-law Tereus, leads to one of the most complex and enigmatic mediatic objects in the *Metamorphoses* (6.576–83): the tapestry in which she weaves purple 'marks' (*notae*) against a white background, sent secretly to her sister, who reads therein the 'pitiful song' (*carmen miserabile*) of all that has happened. This curious textile, which may or may not be a text in the ordinary sense, mimes the sender's body in more ways than one. Specifically as a surrogate for her lost voice, its function has been anticipated in the very scene of her dismemberment. As I have put it elsewhere,

> [F]or a brief, grotesque, but profoundly illuminating moment, Ovid makes us see what *even a tongue* does, usually unheard, while trying to speak. For Philomela's tongue, flung to the ground, tries to crawl back to her, 'murmuring' as it goes. In other words, this tongue briefly becomes a voice, indexical of the body from which it has come; it 'says' nothing, except this: there is a body here, my body.
>
> Butler 2019a: 180

Is this violent tale – easily Ovid's most horrific – somewhere on the mind of Albus Ovidius in his much milder effort to lend human language to bodies that lack a human tongue? It is hard to say: the poem offers no nod to Philomela's myth as told by Ovid or anyone else, unless we find such in the 'red breast' of the poem's swallow, which shares its name with that of Philomela's bloodstained sister, Procne. But the reader who knows Albus Ovidius's namesake cannot really ignore the precedent, not least because it offers some useful principles.[3]

Philomela's tongue, in the Ovidian scene I have just described, demonstrates what we might call the 'vocative principle', from the fact that a word in the vocative denotes the object of direct address but is simultaneously indexical of the voice (*vox*, whence the name) that pronounces it: if I shout your name, you know to look at *me*, either because you recognize my voice or simply because you can locate its source. And what's true of the vocative is true (but not always so obvious) of vocalization generally. I am partly parroting(!) the central thesis of philosopher Adriana Cavarero in her *For More than One Voice*, which she herself expresses at the start of the book in words borrowed from Italo Calvino: 'A voice means this: there is a living person, throat, chest, feelings, who sends into the air this voice, different from all other voices' (Cavarero 2005: 4). I would add only two things to this rule: (1) ancient thought tended to hear 'voices', so called, in nonhuman animals and even in some inanimate objects;[4] and (2) sometimes the voice differentiates not *singulatim* but *generatim* – adult from child, male from female, genus from genus, species from species.

This vocative principle is in evidence right at the poem's start. Who pronounces its apostrophe, that is, its second-person (*tu*) address? Possible answers range from the singular (Albus Ovidius) to the variously generic (the poet, the author, the narrator, the author-function, the reader). Strikingly, the addressee herself similarly oscillates between the singular (Philomela the princess, or a specific nightingale who is the speaker's 'sweet friend') and the generic (all nightingales). Both the oscillation and the mirroring set the stage for what I shall argue is the poem's rather remarkable sonic achievement. After its opening address to the nightingale, the poem turns, as mentioned earlier, to a long catalogue of birds and their songs (followed by a rather more scattershot survey of other animal sounds). Attempts to represent birdsong in ancient poetry go back at least to the Athenian theater, the most famous example being Aristophanes' comedy *The Birds* (Aristophanes 2000), the characters of which includes none other than Philomela's rapist, Tereus, whom the myth likewise transformed into a bird. He first serenades the nightingale (who, in Aristophanes' version, was formerly Procne rather than her sister) and then summons his feathered fellows with a song made partly of bird sounds that are first borrowed back from the onomatopoeic name of his own species (the *epops* or 'hoopoe') but quickly move beyond: *epopopoi popopopi popi, iô iô itô itô itô itô, tiô tiô tiô tiô tiô tiô tiô tiô*, and so on (lines 227–8, 237, on which see Nooter 2019: 200–3). Whatever its date, the poem joins this long tradition, knowingly or unknowingly.

Alongside the literary representation of nonhuman sound, ancient thought developed a complex theoretical apparatus. In particular, the question of the difference between human language and (other) animal sounds later received a highly influential theoretical framing in Stoic thought that was inherited by the late antique grammarians I already have mentioned, who distinguished between the human *vox articulata* and the *vox confusa* of nonhuman sound sources, whether animal or inanimate (discussion and examples in Butler 2015: 113–14; 2019b: 8). The latter, they explain, unlike the former, could not be written down; they then provide examples, like the *mugitus* of cattle. The problem is easy to see (and hear): they've just written down a *moooooo-gitus*. The poem's ornithological catalogue depends overwhelmingly on words like *mugitus* (a word which itself makes an unsurprising appearance later in Albus Ovidius's poem, in the catalogue of quadruped sounds). That is to say, it depends on words that display what linguists call iconicity: signifiers that are *not* arbitrary but that instead actually embody or mime characteristics of their signifieds. (Iconicity is the linguist's term for more or less what literary critic calls onomatopoeia.) The Sausurrean insistence that the sign's pairing of signifier and signified was usually (though not always) arbitrary was leveled against earlier linguistic theories that instead supposed that all language was fundamentally iconic.

The poem quickly overwhelms us with the names of specific bird-species and verbs for their signature sounds. Most (though not all) of the sound-verbs are iconic. For example, the first bird, the *parrus*, which lexicographers have been unable to identify, is given the verb *tinnipet*, i.e., it makes a sound something like *tinnip, tinnip, tinnip*. Other verbs are formed in similar fashion (though it is not always clear where the iconic part ends, and the frequentitive suffix *-it*, underscoring the sound's repetition, sometimes intervenes before the verb ending):

accipiter (hawk)	*pipat*
acredula (unidentified)	*rurirulat*
anas (duck)	*tetrinnit*
anser (goose)	*graccitat*
apis (bee)[5]	*bombilat*
cicada (cicada)	*fritinit*
ciconia (stork)	*glottorat*[6]
corvus (raven)	*crocitat*
gallina (hen)	*cacillat*
gallus (rooster)	*cucurrit*
hirundo (swallow or martin)	*trissat*
merops (bee-eater)	*zinzizulat*
merulus (blackbird)	*zinzitat*
milvus (kite)	*lupit*
palumbes (wood pigeon)	*pausitat*
passer (sparrow or other small bird)	*titiat*
pavo (peacock)	*pulpulat*
perdix (partridge)	*caccabat*
progne (swallow)	*zinzizulat*
sturnus (starling)	*pusitat*
turdus (thrush or similar bird)	*trucilat*
vultur (vulture)	*pulpat*

In several cases, the same sound embedded in the verb can be heard in the name of the bird itself. Thus the *bubo*, probably the horned owl, *bubilat*, i.e., goes *bu bu bu*; the *butio* (a kind of hawk or buzzard) *butit*; the *cuculus* (cuckoo) *cuculat*; the *grus* (crane), to which we shall return later, *gruit*; the *strix* (another kind of owl) *stridit* (the poem applies the same verb to the *vespertilio*, 'bat'); and the *ulula* (yet another kind of owl) *ululat*. (These last two are common verbs with wider application: *stridere* is used for high-pitched sounds from a variety of sources, while *ululare* designates a range of human and nonhuman 'howls'.) The name of the *graculus* (jackdaw) is similarly formed, though the poem gives this bird the verb *fringulit*, one of several related and apparently onomatopoeic Latin words used for broken sounds made by birds and humans (such as crying infants) alike. All these imitations depend partly on the natural pitches of vowels and voiced consonants, or to be more technically accurate, on what linguists call formants. The alphabetic transcription of bird calls and songs is something ornithologists still do today, alongside newer technologies, in part because it is hard to index and search the latter. John Bevis, in *Aaaaw to Zzzzzd: The Words of Birds* (2010), offers a charming introduction to the practice and its principles.

To call such words extra-linguistic transcriptions of nonhuman sounds would only be half right. Nothing is quite so uncanny as picking up a children's book in a language that is not your own and discovering that the animals in it sound somewhat different. In other words, we are taught, even before learning to read, to accommodate the *vox confusa* of the animal kingdom (even in the case of domestic animals) to the phonemes of our mother tongue. Nevertheless, it would surely be wrong to claim that this poem only tells us what the birds it describes sound like *in Latin*. For all you really need to be able to do is to reproduce the basic sounds Latin assigned to the same alphabet you are now reading – taking care, for example, to pronounce *a* as *ah*, as in most modern languages derived from Latin, though matters are messier in English – in order to match some of these words to animals in the wild. Indeed, to suppose otherwise would be to shortchange the poem as an ecological document. Though it might be difficult in practice, it would in theory be possible to try to use these transcribed sounds to locate the author in a particular part of the ancient Mediterranean biosphere, or to gauge the more general diversity of populations of avian species many centuries ago, or even to hear the lost sounds of species now extinct. In the rest of this chapter, however, I want to offer a rather different experiment, predicated on an emphatic shift in perspective. The reasons for this shift will be found in the course of the brief detour that follows.

A stroll with Uexküll

It should go without saying that doubt about the alleged distinctiveness of human beings from (other) animals long predates recent turns to posthumanism and the like. Much of that doubt, new and old alike, can be crudely divided into two categories: the anthropomorphic and the zoomorphic. The former, for example, would comprise all those viral videos of animals behaving adorably like humans, but it would also include the dancing, singing, instrument-playing beasts that frolic in the margins of many a medieval manuscript, as well as the animal divinities that fill the vitrines of anthropological museums. Indeed, nonhuman animals can be found doing characteristically human things almost as far back as the textual and material record can take us. This is true to such an extent that the relatively neat distinction between humans and their predators and prey in early cave painting would seem to be the exception that proves the rule – unless, of course, as in fact seems rather likely, we project onto these an understanding alien to the thought-world of their artists.

The zoomorphic impulse is simply the mirror image of the anthropomorphic one: in other words, a tendency to see humans as embodying the characteristics of nonhumans. The bloodthirsty king Lykaon, in Ovid's *Metamorphoses* (1.216–39), behaves like a savage beast, and so, in the poem's first human metamorphosis, he is transformed into a wolf, lending that species its Greek name (*lukos*). Reversing the direction of this change, to metaphoric and mythological wolf-men we can add the ape-men of Darwinian science, an enterprise that, among other things, catalogued countless ways in which human bodies echo those of living nonhumans and thus of our shared nonhuman ancestors. And further examples of zoomorphic impulses, before, between, and after these two, can be multiplied almost endlessly.

In sum, despite our frequent invocations of human exceptionalism, we humans have long shown a concurrent willingness to regard nonhumans as like ourselves and ourselves as like them, at least in part. A moment's reflection, however, reveals at least one significant limit to this ancient and enduring posthumanism: both of its perspectives, the anthropomorphic and the zoomorphic, are expressed from and as a human point of view. In other words, neither explicitly stops to ask what, if anything, nonhuman animals, left to their own devices, think of us – a question, of course, that risks falling into the pattern it criticizes, anthropomorphizing nonhumans into 'thinkers' about the human. Such a turn is instead central to recent posthumanist work: one thinks, for example, of Donna Haraway in *When Species Meet*, riffing in turn on the naked Jacques Derrida's encounter with the stare of his own cat – as unsettling for Derrida as it surely has been for many of his readers (Haraway 2008: 19–27; Derrida 2008: 3–11). In fact, the inverted perspective has been with us longer than might first seem to be the case. Sometimes, indeed, it can be found quietly at work in the same anthropomorphic and zoomorphic fantasies I have just been describing, further disturbing if not quite dethroning collective confidence in human distinctiveness. In a moment we shall consider the way in which this inversion underpins the *Elegy on the Nightingale*. But first, let us briefly conjure a more recent and less literary attempt to think one's way into a nonhuman animal's point-of-view without pretending to leave one's humanness entirely behind: Jakob von Uexküll's 1934 *Streifzüge durch die Umwelten von Tieren und Menschen*, translated into English in 1957 as *A Stroll Through the Worlds of Animals and Men*, though it has recently been retranslated for a newly enthusiastic post-humanist readership.

Uexküll for a long time was most generally remembered for his emphasis on the term at the centre of his work well before it appeared at the centre of this particular title: *Umwelt*. Today this word ordinarily means 'environment', but in Uexküll's usage it refers emphatically to an environment not as it is surveyed from (or as if from) the outside but, instead, as it is perceived and navigated by a particular organism that inhabits it. And if, for Martin Heidegger, the concept of *Umwelt* makes sense only in terms of anthropocentric ontologies (Heidegger 1962: 84–5, 94–5; helpful discussion in Chien 2006), Uexküll's *Umwelten*, as his title makes plain, are also (and, indeed, primarily) those of nonhumans:

> This little monograph does not claim to point the way to a new science. Perhaps it should be called a stroll into unfamiliar worlds; worlds strange to us but known to other creatures, manifold and varied as the animals themselves. The best time to set out on such an adventure is on a sunny day. The place, a flower-strewn meadow, humming with insects, fluttering with butterflies. Here we may glimpse the worlds of the lowly dwellers of the meadow. To do so, we must first blow, in fancy, a soap bubble around each creature to represent its own world, filled with the perceptions which it alone knows. [. . .] A new world comes into being [. . .,] the world as it appears to the animals themselves, not as it appears to us.
>
> Uexküll 1957: 5

Uexküll's extended essay is deliciously playful; particularly hilarious are its illustrations, by Georg Kriszat, which adopt the point of view of this or that organism on its own 'soap bubble'. Or to be more precise, Kriszat attempts to represent how humans would see a given scene differently if they were attentive to the same things as the organism in question: a living room as mapped by a dog, the sea-floor as perceived by a scallop, a chandelier as seen by a fly (this last illustration relies on a photographic filter). But amidst all the fun is a series of genuine provocations born of great methodological care. Uexküll is keenly aware of the risk of anthropomorphizing the entire animal kingdom; nevertheless, against those who would reduce nonhuman animals to ensembles of mechanistic processes, he insists on the validity of speaking of the animal as a 'subject' (Uexküll 1957: 6, translating German *Subjekt*, Uexküll 1934: 21) even in the case of the lowly tick, his first example. Wherever one decides to locate a mature tick's perceptive capacity, its *Umwelt*, Uexküll argues, is one in which cold-blooded animals are different from mammals, but all mammals, human or otherwise, are equally delicious (or something like that). Indeed, Uexküll will go on to argue that a tick's *Umwelt* even obeys its own particular rules of time.

I do not intend to 'stroll' any farther into the weeds (as it were) of Uexküll's arguments about ticks and the rest. Rather, I simply want to mark how his shift in perspective to the nonhuman 'subject' at once conjures a being who, one way or another, responds to the presence of a human but does not distinguish between its human body and that of any other mammal: all that really matters is the scent of butyric acid released by mammalian skin, the tick's cue to drop from its overhanging branch, to burrow, and to feast. That this cue is an olfactory one is conveniently suggestive of the way in which Kriszat's illustrations translate not only nonhuman *Umwelten* into human terms but also (including as part of that very translation between species) nonvisual into visual perception. Taking my own cue from this sensory drift, I want now to ask what might be gained by asking what humans sound like to nonhumans. To be clear, my goal is not to uncover what that question might tell us about nonhuman animals themselves: that would be a worthy project, but one well beyond my expertise. Rather, my much humbler goal is one that, for better or worse, returns us squarely to human (and even humanistic) matters. In other words, what does imagining a nonhuman perspective on human sounds tell us about those sounds?

I should note that I am not the first to invoke Uexküll along such lines: recently, David Trippett has used him to ask what the 'transhuman ear', by 'virtually extending our *Umwelt*', might do for music, both in theory and in practice (Trippett 2018: 235). 'Viewing the body as upgradable technology', he argues, 'carries the startling corollary that music need not always be conceived and composed according to our biological limits' (Trippett 2018: 242). Trippett, in other words, takes what he also calls the 'extra-human' decidedly in the direction of the more than human, i.e., of transhuman technological enhancement and extension of the human, toward the possible undoing of the very category of the human (i.e., the posthuman). The following comments, by contrast, look not to possible futures, whether utopian or dystopian, but simply around and back at the hearing beings with whom we have always shared a world of overlapping *Umwelten*: namely, nonhuman animals. These, of course, in the traditional understanding, are not more but less than human. But setting aside the (admittedly urgent) question of whether humans should continue to construct such hierarchies at all, the field of sensation already tends to upend their order. Everyone knows, for example, that 'man's best friend' has superhuman powers of both smell and hearing; indeed, Trippett's first musical example includes the 'ultrasonic dog whistles in Per Nørgård's Fifth Symphony (1994)', inaudible to listeners who are merely human (Trippett 2018: 201). Again, however, my interest is not whether cochlear implants or other 'prosthetic auditory technology' will one day enable us to enjoy such works in full (Trippett 2018: 240). Rather, I want to ask how thinking about nonhuman listeners may retrain us – indeed, has long been used to retrain us – in the ways in which we listen to the sounds we already are biologically equipped to hear.

A nightingale listens

With this we return to our poem. As we have seen (and heard), it opens in full mimetic overload, describing the nightingale as 'like no other' precisely in its ability to repeat/ echo (*referre*) the songs it has learned – from other nightingales, other birds, or even from humans. It is, therefore, like the poet himself, who also echoes a 'thousand' (well, not quite a thousand) heard sounds in his own well-modulated song. In other words, the sonorous nightingale is conjured, from the get-go, not just as a music-maker, but as a listener.[7] That listening is itself echoed in the middle of the poem, when the catalogue turns to the parrot, whose bird-sounds, uncannily, consist of two human words for 'hello', one Greek, one Latin, depending, we probably are meant to assume, on the native tongue of its owner: 'The parrot produces human speech with its voice and hails its owner with "*chaere*!" or "*ave*!"' (lines 31–2).[8] At its centre and in its frame, this poem that seemingly is about the various birdsongs of the world quietly invites us to think instead about what the world, including the humans in the world, and even their poems, might sound like to birds. Again, I'm not suggesting that adopting this avian perspective is really possible, at least not fully so. But what might we gain by trying to imagine it?

We might start by noting that, indifferent to words, the nightingale hears this poem as it would any other song. That doubtless seems a birdbrained point to make, but note how it begins to pull our attention away from the specific words with which the poem apes the sounds of the nonhuman world: for the bird, the *whole poem* is made of animal sounds: namely, those of a human. Compare the human reader without Latin or the aid of a translation: the poem is perhaps recognizable as poetry, but not as much else. Conversely, the Latin speaker who also knows Greek may realize that, etymologically speaking, 'Philomela' can either mean 'lover of music' or 'lover of apples', either of which makes good ornithological sense. But in another sense, all of us, including the nightingale itself, hear the same thing, either way.

Letting this avian thinking blend a bit with our human (and humanistic) instincts, just how much is finally human in a sentence like 'Grus gruit in gronnis' (line 23)? Hexter et al. (2020: 53) make a valiant effort to capture the meaning without losing all of the soundplay: 'The crane crunkles in its quagmires'. But the original is fundamentally untranslatable. The crane's sound (*gru gru*) provides both its name (*grus*) and its verb (*gruit*), as was the case, for example, with the *bubo* above. This time, however, even the bird's habitat is described using a rare (and possibly late) Latin word, *gronna* or *grunna*, that clearly means little more than 'a place where you would hear *gru gru*'. In other words, this is a whole world built, populated, and animated by a single nonhuman sound. Still more interesting is the verb at the end of the same line, *drensare*, 'to whoop or honk', describing the sound made by a swan. Here, the human reader with Latin may hear nothing but an arbitrary sign, since it seems at first hard to get a swan-sound out of the verb. But that verb actually derives from an Indo-European root (**dher-*) describing a range of audibly vibratory 'buzzing' sounds – a root seemingly related to another that gave English a word for a bee, 'drone', and Ancient Greek its term for funerary lament, *thrênos* (Watkins 2011: 19). Out of the blue, we suddenly find ourselves making new sense of all those mourning characters in myth, transformed into birds.

My point, to be clear, is not to revive a case for a fundamental iconicity of all language. Rather, I want to use the migration of iconic words through and across languages, through and across species, and even into and out of the environment itself, in order to suggest something more basic about the promiscuity of sound to human ears. Consider the use of *gemere*, in line 20, to describe the sounds of pigeons and doves: 'and the chaste *turtur* (turtle-dove) and *columba* (pigeon, dove) lament (*gemunt*)'. This is a verb with no discernable iconic force. But consider the extraordinary number of sound sources – human and nonhuman, animate and inanimate, natural and manufactured – that comprise its attested subjects. Here are the rubrics within the word's definition in the *Oxford Latin Dictionary* (Glare 1996: 757), minus the examples meant to illustrate them:

1. a. to utter a sound expressing sorrow, pain, regret, etc., groan, moan
   b. (indicating exertion)
2. (of animals or birds)
3. (of things) to give out a deep, hollow, or mournful sound:
   a. (of the seashore, etc.)
   b. (of wooden structures, etc., under strain)
   c. (as a result of being struck by a hard or heavy object)
   d. (of musical instruments)
   e. (of places, with the noise of human activity, animals, etc.)
4. (with acc.) to groan or grieve for, lament

On the one hand, sound here seems to overwhelm the very possibility of differentiation by the sign. On the other hand, I suspect that most of us, reading through this list, sense that these sounds really do belong together, somehow. This is despite the fact that English does not map out sounds along exactly the same lines: striking is the number of sub-meanings for which the dictionary offers no specific English equivalent, providing instead, in brackets, the various categories of sound objects to which the word is applied. Nevertheless, this Latin *Umwelt* still strikes a chord.

More than a few poetic movements – one thinks first but hardly only of Romanticism – have regarded such a chord as the heartbeat and lifeblood of poetry itself. But as a rule, such declarations depend on the supposition that poetry blurs what language has differentiated. Our forgotten Latin poem, on the contrary, suggests something rather more radical: namely, that language itself, despite itself, always already makes us and our world sound remarkably alike. Likeness, of course, is the poem's framing conceit: the poet-narrator proposes to match the nightingale's own ability to imitate any 'voice' (*vox*) or 'tune' (*modus*) it hears. This means that all of his single acts of animal mimesis add up to an imitation of the pan-mimetic nightingale itself. What will his 'sweet friend' then do when she hears his song? Naturally, she will sing it back to him. Who, we may ask, is finally echoing whom? It does not really matter: these two friends are one another's *alter(a) ego*, each an *alterius echo*, an 'echo of the other'. All manner of difference seems to collapse in this echo chamber: between human and nonhuman, male and female, this species and that species. This elision of difference, however, is a far cry from the attempted annihilation of the Other by Philomela's rapist in the nightingale's etiological myth. For this poem is nothing if not a celebration of difference, in the guise of sonic biodiversity; its unity is forged very much *e pluribus*, like that of the 'great animal orchestra' of soundscape ecologist Bernie Krause (2012), or what Dominic Pettman (2017: 7, 65) calls 'the ecological voice' or *vox mundi*. Strikingly, the poem's possibly tacked-on closing distich devolves this unity upon a monotheistic deity, the only listener capable of hearing in full the vast and varied ensemble that includes, we are told, even the world's silences. Note, however, that this finale characterizes that god in terms both anthropomorphic and zoomorphic: the superhuman listener is like a superpowered nightingale.

It is here that the promiscuity of sound, buzzing through all manner of media, meets the material conditions of expression: human, avian, and otherwise. Our bodies vibrate with sounds that are never entirely our own. Even the groans that seem to come from the cores of those bodies and from the depths of our pain are borrowed from and lent back to a resonant world. In this regard, the poem's loss of any authorial inscription seems the culmination of its poetic project, in that the author has been entirely eclipsed by his counterpart, the nightingale, itself an echo of all it hears. Who wrote this poem? The poem answers, 'My sweet friend . . .'

Notes

1. This chapter is a revision and expansion of Shane Butler, 'Animal Listening,' *Journal of International Voice Studies* 6.1 (2021): 27–38, with the kind permission of the journal. Earlier versions were presented at the University of Helsinki, at the Max-Planck-Institut für Wissenschaftsgeschichte in Berlin, at Cornell University, and at the annual meeting of the American Comparative Literature Association. Heartfelt thanks are extended to organizers and audiences on all those occasions, as well as to the editors of the present volume.
2. I shall follow their Latin text, which is based on that of Klopsch 1973, though my translations are my own.
3. On the long reach of Philomela's 'sob' in literary deployments of the nightingale, see Connor 2014: 64–7.
4. On the range of meanings of the principal Greek and Latin words for 'voice' (respectively, *phônē* and *vox*), see Butler 2015: 112–15; Butler 2019b: 4–8.
5. In the case of the bee and the cicada (following), being winged and famously sonorous evidently sufficed to win inclusion in this otherwise avian symphony.
6. While this verb clearly works by onomatopoeia, it may also depend on the Greek word for 'tongue', *glôttis*, which is also given by Pliny, *Natural History* 10.67 as the name of one species of bird with an especially long tongue.
7. Compare LeVen (2019) on 'creative listening' by the 'erogenous ear', including that of Ovid's Echo, revised and expanded in LeVen (2020: 107–35), where it is accompanied by a chapter on some of classical literature's nightingales (168–205).
8. Hexter et al. (2020: 53) take the distinction as being, not between languages, but between 'hello' and 'goodbye', but this stretches the Latin while weakening the mimetic point.

CHAPTER 4
MULTISPECIES TEMPORALITIES AND ROMAN *FASTI* IN OVID'S *METAMORPHOSES*

Francesca Martelli

Among the objections to the term Anthropocene that Donna Haraway lists in *Staying with the Trouble* is a particular set of doubts about the narrative scope of the concept it describes: 'The myth system associated with Anthropos is a set-up, and the stories end badly', she writes. 'It is hard to tell a good story with such a bad actor. Bad actors need a story, but not the whole story.'[1] The complaint about making Anthropos the protagonist of this story of deep ecological time is a familiar one: too broad a category to convey accurately which humans are responsible for the planet's current state of ecocide,[2] it is at the same time too narrow a designation, enshrining the human at the centre of things, a position that various currents in the humanities and social sciences have for some time now contested. But Haraway's chief emphasis here is on a less familiar issue – that is, the subsidiary problem of narrative scope that arises as a consequence of placing the human at the centre of this story. Her implication is that the totalizing narratives that tend to get told under the sign of the Anthropocene are symptomatic of a tendency on the part of humans to view the planet and its crises at a distance: they assume that we have a complete, godlike purview of the narrative of the earth's dereliction, and that we can see the consequences of our disastrous actions *in toto*, as if we were capable not only of gazing on the planet from afar but of situating ourselves outside of time as well.[3] Multispecies storytelling is the alternative narrative mode that Haraway and others advocate for this time of planetary ruin: born out of ethnography's interest in culturally situated relationships and interactions, its stories tell of the contingent encounters that take place between species from within the ecologies and economies in which they are enmeshed.[4] The perspectives of such stories are always partial, constrained as they are by the located nature of the particular encounter involved, as well as by the limitations posed by translating difference across species barriers.[5] Multispecies storytelling is, then, the narrative antidote to the Anthropocene's totalizing visions.

Ovid's *Metamorphoses* invites readers to map its narrative prerogatives, and disavowals, onto a tension that resembles the distinction between these alternative modes of planetary storytelling. Its epic status promises a totalizing vision, one that we expect to find focusing its narrative universe around the goals of a particular community of human actors. Yet this promise is immediately renounced, as the poem eschews any focus on a singular human protagonist in order to resituate the human in relation to a diverse array of other life forms, whose ancient ethnographies are incorporated into the narrative at the moment of one species' transformation into another. A significant effect of the expansion of perspective that we gain from this poem's multispecies cast of protagonists, is a concomitant dilution of its totalizing goals. For while on the face of it, the *Metamorphoses* outdoes its epic predecessors in narrative scope by taking in the whole of cosmic history, from the beginning of time until the moment of Ovid's writing, in an exaggerated version of universal history, it is hardly a totalizing story. It offers no coherent overview of the times that it covers, no clearly discernible divisions or transitions in its organization of historical time, and no account of how or why things progress as they do. It proceeds, rather, from one story to the next by means of contingent and opportunistic narrative connections,[6] wandering sideways across the chance encounters of different beings, its only unifying thread being the theme of change in time, much like the stories of multispecies ongoingness that Haraway charts in *Staying with the Trouble*.[7]

Underpinning Ovid's narrative agenda was a cultural revolution that had come to centre on the control and organization of time. Augustus, the first Roman emperor, had identified himself as history's heir by imprinting himself on two of the most important time-keeping devices at Rome: firstly by implementing the calendar reforms initiated by Julius Caesar to align the annual Roman calendar, the *Fasti Annales*, with the solar year; and secondly, by modifying the so-called consular *Fasti*, Rome's chronicle of linear, historical time, so that it now anchored the history of the city not only to the dawn of the republic, by naming each year since then after the two consuls for that year, but also to the founding of Rome by the legendary Romulus, Augustus' pre-republican alter ego.[8] Scholars usually see the new emperor's two-pronged assault on linear and cyclical axes of time as an important spur for the two major poetic projects of Ovid's middle career, the *Metamorphoses* and the *Fasti*, Ovid's versification of the Roman calendar. The uneven chronological progress of the *Metamorphoses*, which seems to move backwards and sideways more obviously than it ever moves forwards, which contains a number of pointed anachronisms, and which bypasses Augustus to end with Ovid's own apotheosis in fame, have all been taken as part of the poet's challenge to the emperor's attempt to make the arrow of historical time point at him.[9]

But another reason for the poem's complicated chronological progress is that it incorporates the temporal rhythms and histories of a variety of other life forms into its narrative. Locating the linearity of human (more specifically, Roman) time may be very difficult when it is intersected by the temporalities of many other species, whose lineages find expression in patterns of cyclical time. With these patterns, the other-than-human organisms that populate Ovid's poem also complicate the neat separation of linear and cyclical time that seem to be shared out between the *Metamorphoses* and the *Fasti*, and show how the web of zoological life might confound human, and more specifically imperial, attempts to organize time into these discrete categories. One consequence of multispecies storytelling, Ovid shows us, is that the respective temporalities that different species embody and articulate in relation to one another warp any sense of a singular linear timeline altogether, as linear time is replaced by multiply intersecting lineages; and that these lineages are themselves constituted by the transmission of annual cycles of behaviour.

In his study of the narratives of extinction currently facing a number of endangered bird species, Thom Van Dooren offers a conception of species that is premised directly on the repetition of such annual cycles. Van Dooren uses the term 'flight ways' as a means of moving away from static taxonomic categories and of attempting to define species instead by situating them in time. According to this view the properties that define a species are not a set of fixed biological criteria, but are rather an evolving set of behaviours: patterns of migration, rituals of courtship and mourning, and all the other morphological and behavioural characteristics that are handed down between generations in their cycles of living and dying. The Darwinian premise for his insight is that a species is, in his words, 'a line of movement through evolutionary time'.[10] But if that makes the temporal trajectories of flight ways sound abstract, they are in fact anything but. For underpinning this view of species is a notion that Van Dooren borrows from Deborah Bird Rose of 'embodied temporality', which is to say the way in which each member of a species embodies in the flight ways that take up its brief lifespan an intergenerational inheritance that stretches back many millions of years and which stretches forward for an unknown period of time as well.[11] Each member of the species is thus a form of time capsule. Or, as Van Dooren puts it, 'any individual bird is a single knot in an emergent lineage, a vital point of connection between generations – generations that do not just happen but which must be achieved.'[12]

While the evolutionary premises of Van Dooren's conception of species, as well as the imminent threat of biodiversity loss, may have been alien ideas to Ovid, his narratives of multispecies encounter display a kindred interest in identifying different life forms by exploring the temporalities that these forms embody at the intersection of linear and cyclical time. In this chapter, I want to draw on two examples of this that receive particular emphasis in the poem, and which happen to deal with types of bird, in order to elucidate the various 'knots of time' that tie the birds in question both to the lineages of their own species and also to others within the wider zoological web. While the evolutionary content of Van Dooren's flight ways may be missing from these vignettes of typological behaviour, his description of the lineages that may be expressed in the reperformance of a particular routine action strikes a chord with them. In particular, because the rituals that Ovid picks out for the birds in these two episodes deal with their annual habits of breeding and dying, they focus our attention on the question of the survival and perpetuation of the larger species to which the birds belong, and invite us to appreciate the intergenerational persistence of the species concerned as an achievement, in and of itself. For Van Dooren and other contemporary multispecies ethnographers who focus on narratives of extinction, the stories that they tell about various species have become narratable because those species have come (or are coming) to an end. For Ovid, who had less cause to dwell on questions of extinction, it is the origin stories of different species that is of interest.

Halcyon generation(s)

In Book 11 of the *Metamorphoses*, at the end of the story of Ceyx and Alcyone, this happily married couple are torn apart when the husband, Ceyx, is killed in a sea storm, only to be reunited when they are subsequently turned into halcyon birds, a species that is commonly identified as a kingfisher. Ovid's rendition of this story stands out for its amplitude (as one of the longest episodes in the poem),[13] and for its contribution to the poem's discourses of desire (as one of very few stories of mutual desire between spouses).[14] Equally striking is the description of the birds that the married couple become after their metamorphosis. The halcyon bird was the subject of a wealth of mythographical and pseudo-scientific lore in the ancient world, much of which Ovid incorporates (and comments on) in his own ethnography of the bird that Alcyone becomes (*Met.* 11.734–48):

> Dumque volat, maesto similem plenumque querelae
> ora dedere sonum tenui crepitantia rostro. 735
> ut vero tetigit mutum et sine sanguine corpus,
> dilectos artus amplexa recentibus alis
> frigida nequiquam duro dedit oscula rostro.
> senserit hoc Ceyx an vultum motibus undae
> tollere sit visus, populus dubitabat; at ille 740
> senserat, et, tandem superis miserantibus, ambo
> alite *mutantur*. fatis obnoxius isdem
> tum quoque mansit amor, nec coniugale solutum est
> foedus in alitibus; *coeunt fiuntque* parentes,
> perque dies placidos hiberno tempore septem 745
> *incubat* Alcyone pendentibus aequore nidis.
> tum *iacet* unda maris; ventos *custodit* et *arcet*
> Aeolos egressu *praestatque* nepotibus aequor.

> And as she flew, her mouth, by now a slender beak, gave forth such sounds as seemed to come from one who knew lament and grief. And when she reached the silent, bloodless body, enfolding his dear limbs with her new wings, she tried to kiss him, but in vain – her beak was hard, her kisses cold. Whether Ceyx felt this, or whether he just seemed to lift his face with the motion of the waves, people were unsure. But he had felt it. And at last, through the pity of the gods, the two are changed into birds. Having suffered the same fate, even then their love remained, nor was their marriage vow sundered by their winged form. They mate and breed; and for seven peaceful days in the winter season Alcyone broods upon her nest as it floats upon the sea. At this time, the waves of the sea lie still; for Aeolus guards his winds and forbids their escape, and keeps the sea calm for the sake of his grandchildren.

At lines 734–5, Ovid's halcyon emits mournful noises, a characteristic of this bird first attested in Homer.[15] In lamenting and then embracing Ceyx, Ovid's halcyon displays traces of a tradition that maintained the conspicuous devotion of the female halcyon for her mate, a devotion that is perhaps best illustrated by the behaviour described by Aelian, in which we hear how the halcyons, which is the name for the female members of the species, would customarily carry their mates (the male *keruloi*) on their backs when they were too old to fly.[16] But above all, Ovid's newly formed halcyon bird displays the unusual breeding habits that are so distinctive a feature of this bird in ancient lore: in line with a tradition that predates Aristotle, but to which Aristotle also seems to subscribe, Ovid's halcyon breeds in mid-winter, building its nest on the surface of the sea during the halcyon days, the days around the winter solstice when there was said to be a brief respite of fine, calm weather amid the otherwise prevailing winter storms.[17]

Here in this annual breeding routine we find a quite distinctive flight way: a pattern of behaviour that these birds will bequeath to their descendants, who will mark their connection to their halcyon ancestors in their performance of this annual ritual. Particularly striking in Ovid's description of this annual breeding custom is the tense of the verbs used to describe it: the story of Ceyx and Alcyone is told in the preterite up until the point of Ceyx's metamorphosis, when the halcyon bird's devotion to its mate and annual breeding custom are described in a series of present tense verbs (*coeunt fiuntque ... incubat*). These are historic presents, which describe the behaviour of the original halcyon bird, aeons ago. But they are at the same time continuous presents, which describe the ongoing (and unfinished) actions of her avian descendants, who repeat the same action, year after year, even at the time of this poem's narrating moment. The tense of these verbs conveys the potential expansion of time that may be held open in a historic present, as Alcyone's annual breeding ritual connects her metonymically to the iterations of that ritual that each subsequent generation of her species will perform, in ways that underscore the individual bird's capacity to compress the linear time of her species. This capacity is further underscored by the ambiguity of the reference to Aeolus' 'grandchildren' (*nepotibus*) at *Met.* 11.748, which could refer specifically to Alcyone's immediate progeny, or metonymically to all the generations that succeed them.

But there is a stronger sense in which this particular habit casts a spotlight on the intergenerational connections that link the past and future of the species. One of Van Dooren's extinction narratives focuses on the current plight of the Laysan albatross, an ancient bird, thought to predate anything that resembles a human by millions of years, that is currently imperilled due to the location of its breeding ground on Midway Atoll, in close proximity to the North Pacific Garbage Patch, where its ingestion of plastics and other related toxins has had catastrophic consequences on its capacity to breed and flourish.[18] A particular effect that the ingestion of these toxic plastics has had on this bird has been to thin the shells of the eggs that it lays, so that they break prematurely and the young chicks die before they are properly hatched. Van Dooren highlights the poignant coincidence in the fact that the impact of these substances is felt almost entirely by breeding birds and their young – at the precise point where one generation brings forth the next, as if to highlight the fragility of the intergenerational achievement that is the albatross flight way.[19]

The story of Ceyx and Alcyone, and the halcyon bird that they generate via metamorphosis, may not be told under the threat of extinction, but it too emphasises the precariousness of the moment of breeding and birth, as if to highlight the broader significance of Van Dooren's point for many other species' lineages. Ovid's elaborate rendition of the story behind this metamorphosis, with its emphasis on the dangers of the sea, can be read in retrospect as an *aetion* for this particular ethnographical detail – not the halcyon bird *per se*, but its distinctive breeding season and counterintuitive habit of entrusting its eggs and fledgling young to the winter sea. The miraculous period of calm that is said to descend during the so-called halcyon days makes the sea itself appear to be atoning for its killing of Ceyx by bestowing this favour on his avian descendants, but it is hardly reassuring. If this moment in the bird's annual cycle is one that binds the future life of the species to its past, it is a supremely fragile one, one that emphasises how precarious the lineage of this species is. In dwelling on such moments, Ovid suspends the relentless narrative of linear human time in order to show how it intersects with flight ways of such comparable tenuousness.

But species do not live in isolation from one another; they synchronize their own cycles of behaviour with the life cycles of other species on which they rely for sustenance, in what Deborah Bird Rose calls 'knots of embodied time'.[20] The Laysan albatross is knotted in time to the breeding patterns of the fish that it eats, and has become disastrously knotted to the plastics that have transformed its habitat. So too Ovid's story of the halcyon bird weaves together a number of other avian encounters that speak to the temporal entanglement of its flight ways with those of other species. Ceyx's fateful journey across the sea is driven by a desire to discover the true fate of his cruel brother, Daedalion, who is rumoured to have been transformed into a hawk, a bird that is specifically described as predating on other birds, in a metamorphosis myth that Ovid is thought to have invented.[21] At the end of their story, an anonymous old man watching the halcyon birds relates the story of another waterbird, the merganser, which is said to be a transformed Trojan prince.[22] This story ends with a description of the merganser's distinctive habit of diving down below the water only to resurface again, a behaviour that commemorates the lovesick suicide attempt that its human forebear undertook in throwing himself off a cliff into the sea, only to be saved in mid-fall by metamorphosis. In this weaving together of different bird transformations, Ovid's narrative replicates the flight ways of different bird species as an interlocking system, in which the birds are knotted not just to their own species in diachronic time, but also in synchronic time to each other.

What might such a system look like in light of the ancient lore that surrounds these birds? Something like this, perhaps: the hawk is a predator that would eat a kingfisher, which might be assumed to build its nest at sea in order to protect its young from the predations of land birds like this. And even though Ovid never says so, it is possible to imagine the halcyon building its nest out of the fishbones that the merganser bird spits out after diving down below the water to feed. This reconstruction is speculative, to be sure. But because the narrative is organized as a system of interlocking stories, it lends itself to being decoded in these systemic, ecological terms, even if Ovid leaves the burden of imagining the precise details of the interactions to the reader. In this way, the *Metamorphoses* engages its readers in a form of speculative fabulation, the mode of storytelling that Haraway insists we need to develop in order to imagine our multispecies world – both as it is currently, and differently for the future.[23]

But the more we press the story of the halcyon, the more evident it becomes that almost the entirety of her ancient ethnography is already a speculative fabulation, invented by humans to answer to their own imaginative needs. This much was made clear by the zoological inquiries of D'Arcy Wentworth Thompson, whose *Glossary of Greek Birds*, published in 1895, set out to identify all the species of bird mentioned in Aristotle's *Historia Animalium* against actual species that were known to exist. Under the entry for halcyon, he confirms what modern readers may have suspected: that the elaborate ethnographical lore that the ancient Greeks and Romans developed around the halcyon bird bears no correspondence to the known behaviour of any kingfisher or indeed to that of any other bird that we can identify.[24]

This demystification of the halcyon's flight ways reveals the investment that ancient writers placed in imagining them (whether or not they actually believed in them).[25] Not just in their idealization of the birds' devotion toward one another as a foil for their disappointment about human behaviour; but also in the idea that the harshness of winter might be suspended at the height of the season in order to fledge the young of this largely imaginary bird. The halcyon days map onto Roman calendar time, because they are known to span the winter solstice, a time of celebration in many cultures. In Rome, they coincide with the Saturnalia, one of Rome's most important holidays, characterised as a time of social inversions, when masters waited on slaves, and various other social rules and norms were suspended. The mythical season of the halcyon days provides a counterpart in the domain of natural history to the suspension of norms that we see at this time of year in human culture, and demonstrates the extent to which ancient Romans relied on other species, however imaginary, in order to conceive of, and take, a holiday from themselves.

Mourning Memnonides

If Ovid's account of the halcyon birds calls our attention to the way in which the temporality of a particular life form may be articulated through the birthing practices that it hands down from one generation to the next, another example of typological behaviour that he elaborates in substantial detail focuses, neatly enough, on death and mourning. Mourning rituals are also important for the articulation of intergenerational time, even if they deal not with the organic life of a *genos*, but with its social life, and have a different way of relating the future to the past.[26] In one of the briefest episodes in Book 13 of the *Metamorphoses*, Aurora, goddess of Dawn approaches Jupiter, to request some form of compensation for the death of her son Memnon, a mortal recently killed by Achilles. Jupiter assents by creating a new species of bird, the Memnonides, which are distinguished by an elaborate mourning ritual that they are said to perform year after year: they fight each other to the death, an action that has led a number of commentators, including Thompson, to identify this bird species as what modern readers know as the Ruff (or *Philomachus Pugnax*), on the basis of its characteristic aggression. But where our Ruff displays its aggression primarily in the mating season and is observed in male birds, Ovid follows an ancient tradition that associated the aggressive displays of the Memnon bird with a mourning ritual performed to honour the hero Memnon at the site of his tomb at Troy. Like a number of other mourning birds in the *Metamorphoses*, Ovid's Memnon birds are also female (*Met.* 13.600–19):

> Iuppiter adnuerat, cum Memnonis arduus alto 600
> corruit igne rogus, nigrique volumina fumi
> infecere diem, veluti cum flumina natas
> exhalant nebulas, nec sol admittitur infra;
> atra favilla volat glomerataque corpus in unum
> densetur faciemque capit sumitque calorem 605
> atque animam ex igni; levitas sua praebuit alas.
> et primo similis volucri, mox vera volucris
> insonuit pennis; pariter sonuere sorores
> innumerae, quibus est eadem natalis origo,
> terque rogum *lustrant*, et consonus exit in auras 610
> ter plangor; quarto seducunt castra volatu.
> tum duo diversa populi de parte feroces
> bella gerunt rostrisque et aduncis unguibus iras
> exercent alasque adversaque pectora lassant,
> inferiaeque cadunt cineri cognata sepulto 615
> corpora seque viro forti meminere creatas.
> praepetibus subitis nomen facit auctor: ab illo
> Memnonides dictae, cum sol duodena peregit
> signa, *parentali* moriturae more rebellant.

> Jupiter nodded, and Memnon's steep pyre fell in the high flames, columns of black smoke darkened the day, as when rivers emit clouds of vapour, formed from their own water, and the sun can't break through; black ash flies, and rolled into one mass, it thickens, and takes form, and draws heat and life from the fire; its own lightness gives it wings. At first like a bird, but soon enough a real bird, it whirred on its wings; at the same time clamoured numberless sister-birds born from the same source. Three times they circle Memnon's pyre, and three cries in unison rise up into the air; at the fourth circle they divide their forces. Then two fierce groups on different sides wage war, they vent their anger with beaks and hooked claws, and batter the chests of their enemies. Their bodies, born from the ash of the buried man, fall as death offerings, and they recall that they were created from a brave hero. Their author gives a name to these winged birds: they are called Memnonides after him, and when the sun completes its journey through the twelve signs, they resume their fighting, destined to die in the manner of their father.

This is an extraordinary episode from the point of view of multispecies relations, not least because its depiction of an avian mourning ritual contests a common argument made in favour of human exceptionalism that one way in which the human is distinguished from other species is in its capacity to know death.[27] This is a point that Derrida among others disputes on the grounds that many species of non-human animal perform observable acts of grieving.[28] Ovid would seemingly agree with Derrida, whether or not he has in this instance misidentified a mating ritual for a mourning one, and even if his own take on this question contains a further multispecies twist. For the ritual that his birds perform honours Memnon, who is both the ancestor (*auctor*) of their species, but who is also a human. Rather than effacing the troubling implications of this paradox, Ovid chooses to highlight them. For in fighting each other to the death, the birds replicate Memnon's final death scene, his fatal duel with Achilles, at the same time as they also fall as sacrificial victims to honour the dead hero. But if this would seem to reinforce traditional hierarchies of being that the ancient world observes, underpinned as it is by animal sacrifice, these hierarchies are overturned in the course of Ovid's narrative; and the poet ends up using this myth as an opportunity to explore the variety of possible ways in which species barriers can be made to break down.

In his reading of the cross-species identifications explored in *The Silence of the Lambs*, Cary Wolfe describes how the law of culture, underpinned by the regime of carnivorous sacrifice that is essential to the traditional hierarchy of subjects in the West, arranges species significations on a kind of grid:[29] we move from animalized animals at one end of the scale (the sacrificial animals on which the assumptions of speciesism depend), through to humanized animals (pets, essentially, which are exempt from the sacrificial regime), from there to animalized humans and finally to the category of humanized humans, sovereign subjects, which Wolfe describes as an example of 'wishful thinking'. According to Wolfe's schema, between the sacrificial animal and the hypothetical sovereign human there are graded shades of difference, and these differences serve to expose the ideological fictions that sustain the polar categories (the animalized animal at one end, and the humanised human at the other). How might Wolfe's grid work for Ovid's Memnon episode?

At first sight, the mourning rites that the Memnon birds perform for the dead man, would seem to cast the birds on the lowest rung of species significations: sacrificial animals, dying for the human, who is here already dead, as if to highlight the symbolic surplus of their sacrificial act.[30] But the form of their ritual fighting (line 612 f.) complicates this identification: it looks familiar to a Roman reader because it resembles gladiatorial combat, a ritual form of violence originally associated with the Roman funeral, where the family of the deceased would hire human gladiators to honour the dead by attempting to kill each other.[31] To some extent this humanizes the birds, not because it exempts them from the sacrificial regime, but because Roman ritual, which includes human sacrifice in the form of gladiatorial combat, complicates the grid of species significations as Wolfe presents it. Here human actors, gladiators, provide the paradigm of violent combat and are used to convey the ferocity of the birds' fighting, rather than the other way around. But if the birds are humanized by this logic, this is again complicated by the status of the gladiator himself in Roman thought, who, if not entirely assimilated to the category of the animal, is certainly subhuman, equated as he is with the most degraded rank of slave.[32] Ovid's description of the ritual mourning that the birds perform takes us through this dance of different species distinctions, from the animalized to the humanized animal, and thence to the animalized, or at least de-humanized, human. All this to assert the humanized humanness of Memnon. Or rather to compensate for it, since his humanity is presented to us from the very outset of the story as a form of impoverishment, the privation of divinity. Reading this passage with Wolfe's grid of species distinctions, helps us to see how exactly the distinctions between human and animal start to slide, even in a situation where the ostensible hierarchy between them is overtly premised on the logic of sacrifice.

But if species distinctions break down in this episode through relations of resemblance, they are also transcended by being organised on a genealogical axis as well. Memnon is described as the parent of this species; and despite the clear fact of his biological difference from them, Ovid takes particular pains in this episode to overcome this difference and provide an imaginable (if not scientifically plausible) material basis for their genetic relationship. The birds are not the direct outcome of Memnon's metamorphosis – he does not transform directly into a ruff – as is the pattern for most of the metamorphoses that transpire in the course of the poem. In this story, the alterity of Memnon's humanness is neutralized when he dies, is cremated and transformed into a material substrate, ash, out of which the birds are subsequently born. This process also gives a visually plausible basis for the metamorphosis, which is accounted for at a phenomenal level before it actually materializes, with the description of the ashen shapes in the sky. It feels like the barest ontological shift for these shapes to become the beings that they resemble – another way in which Ovid minimises the distinction that separates this brand-new species of bird from its human origin.

What implications do these cross-species identifications and slippages hold for the status of the ritual as a recurrent annual event? The birds are said to perform this act of ritual self- sacrifice on an annual basis at the site of Memnon's tomb, a ritual that would seem to enshrine their position of subordination in the traditional hierarchy of species. Yet, as if to complement the intersubjectivity of humans and birds in this passage, Ovid implicates the avian ritual in a history of human mourning, by making it serve as an *aetion* for the festival of the *Parentalia*, the annual festival in the Roman calendar when Romans would go to pay their respects to their ancestral dead. Before they embark on their ritual of mutual slaughter, the birds seem to be performing the very human rituals associated with this funerary festival: they encircle the pyre three times in what looks like an elaborate performance of a *lustratio*, the purification ceremony that took place at Roman funerals, and annually at the *Parentalia*, where families would process around the tomb of their deceased ancestors, making offerings. Gladiatorial fighting is also thought to have featured regularly at this festival. And Ovid finally identifies this scene's association with the *Parentalia*, with the adjective *parentali* used to describe the manner in which the birds renew their fighting each year, which could refer either to Memnon (their parent), or to the ancestral claims of the festival.[33] By inventing a link between the ruff's imagined mourning ritual and the *Parentalia*, Ovid encourages us to see the other-than-human aspects of the Roman rituals performed on this day: the encircling of the tomb, which has its civic counterpart in the encircling of the *pomerium* (the outer limit of the city) at times of crisis, looks like precisely the kind of territorial behaviour that is more easily observed in other species than in our own.

The episode ends, though, by eschewing interspecies substitutions and the sacrificial logic that underpins them altogether, and suggests another way of relating across species in calendar time (*Met.* 13.620–2):

> ergo aliis latrasse Dymantida flebile visum est;
> luctibus est Aurora suis intenta piasque
> nunc quoque dat lacrimas et tot**o rora**t in orbe.

> And so, while it seemed pitiable to the other gods that Hecuba had been condemned to bark, Aurora was consumed with her own grief, and still now weeps tender tears and sheds dew on the whole world.

In the closing lines of the episode, the birds' annual sacrifice is presented as inadequate compensation to Aurora for her loss, and another mode of recurrent temporal marker is introduced as an alternative sign of the goddess' grief. Aurora weeps, and bedews the world with her tears – an event, we are told, that still happens to this day. Again, linear time collapses, and we are invited to read the entire episode in retrospect as an aetiology for the daily occurrence of dewfall that marks the appearance of dawn. This too is a commemorative marker of calendar time, albeit one that outdoes the birds' ritual in both frequency and universal reach: all land-borne species (at a certain elevation) experience dewfall on daily basis. And it too presents a moment of cross-species confusion, albeit one that promotes sympathy for other categories of being, rather than their sacrifice: because shedding tears is not something gods and goddesses are supposed to do in Graeco-Roman myth,[34] Ovid's Aurora crosses a form of species barrier by crying,[35] and thereby demonstrates a degree of sympathy for humanity that goes beyond the kind of feeling that tears might normally be held to express.

The differences between these two commemorative actions are quite literally spelled out for us by the ways in which Ovid inscribes the names of Memnon and Aurora into each of them, for both actions explore the commemorative function of names. The interspecies substitutions of the bird ritual are perfectly expressed by the transfer of Memnon's name from himself, the human, to the birds that die in his stead. If the sacrifice expresses one mode of substitution, then the proper name, governed like all linguistic signs by convention, expresses another – and here the two are made to merge, when the birds are given Memnon's name to wear (*praepetibus subitis nomen facit auctor*). Aurora's signature operates quite differently. At the end of the passage, her name appears as a pun in the closing phrase that describes dew falling throughout the world (*toto rorat in orbe*).[36] The way in which the letters of her name are redistributed in the different words that make up this image suggest a transformation taking place, not across different species, but within Aurora herself – or rather between the goddess and the material symptoms of her grief. This seeping cryptogram makes Aurora seem to dissolve into the dew that inscribes her presence (and her grief) throughout the world each morning. But while dew may be a symptom of Aurora's grief, it is a sustaining benefit for many plants and the eco-systems that they support. This is a metamorphosis that affects other species not as a form of sacrificial privation, but as a gift.

Conclusion

The *Metamorphoses* moves forward in time from the beginning of the world to the author's moment of writing, and beyond it to the predicted moment of the poet's fame, which is, among other times, our own present moment of reading. On the way, we are told that this will also be a time that we will share with a number of plants and creatures, whose stories of origin take up much of the poem's narrative space. And so we do: the ruff and the kingfisher (if, indeed, the halcyon is a kingfisher) are still around today, and seem to be coping well enough with the challenges posed by climate change and other forms of planetary destruction. In the particular episodes on which I have concentrated we get the strong version of this insight about sharing time with other species. The routine actions of these particular birds, which express the lineages of their distinct species, are mapped onto the Roman calendar and serve as reminders of the extent to which human temporalities intersect with those of other life forms. As Roman readers live through the year, these moments in calendar time will remind them of the extent to which the rituals that they use to mark the year's movement resemble those of their (and our) feathered kin, and provide an opportunity to reflect on what we share with and owe to them.

But what do we do with the time in between their origin stories, and the metaleptic now of the narrating present? When Ovid tells us that the halcyon birds still breed as they did, is he implying that linear time stands still for the birds as a species? Is he using ancient ideas about species to disrupt the chronological movement of the poem? For Aristotle, for example, species are eternal:[37] since organic integration and reproduction always accompany one another, the function of the latter must be to ensure continuity of form in a species from generation to generation. So while individual organisms may perish, the form (*eidos*) that they inherit and pass on is eternal. But Ovid has already distanced his account of species from Aristotle's by giving his creatures stories of origin. Whether or not we are meant to infer a finitude for these species is not spelled out.[38] But simply by telling stories of their origin, and setting them against one another in the history of the wider world, the *Metamorphoses* takes an important step toward situating species in time, and in this respect represents a significant imaginative bridge somewhere between Aristotle on the one hand, and Darwin on the other.

More productive an approach might be to see the suspended present that Ovid uses to describe the behaviours of these and other animals as a comment not so much about species in the abstract but about the experience of multispecies temporalities – that is, the lived experience of sharing time with different kinds of being. The idea of linear time that the narrative structure of this poem presupposes in its movement from the origins of the world to Ovid's day becomes visible only when we isolate the human from the assemblage of other species with which it intersects. When we look at a wider segment of that assemblage, time invariably sprawls sideways, its forward movement indiscernible amid the focus on what other species are doing at the same point (more or less) in the annual solar orbit, and on how this connects them to their ancestral pasts. These lineages intersect the lineages of human time, but are at the same time incommensurable with them, such that when we see their points of intersection, the poem's narrative tempo changes. Multispecies ethnographies describe queer ecologies, in which beings relate to one another across species barriers in non-reproductive ways.[39] The temporalities that these ethnographies produce are likewise steered – and queered – away from the linear, generational pattern that these stories would tell if they were limited to their own species.

Notes

1. Haraway 2016: 49.
2. As Emmett and Nye point out (2017: 104), 'few scholars of the global South have embraced the concept of the Anthropocene.' See Chakrabarty 2009: 216 for a (brief) postcolonial critique of the Anthropocene's imperative to view human agency (and responsibility) for human-induced climate-change as belonging to the species as a whole, rather than to the members of a smaller group of wealthy nations. This critique leads some scholars, e.g. Moore 2016, to prefer the term Capitalocene.
3. Latour 2017: 110–45 connects the totalizing – and distancing – visions generated by the Anthropocene with the image of the Globe in the history of Western thought. Cf. esp. Latour 2017: 127: 'we realize that the Globe is not that of which the world is made but rather, a Platonic obsession transferred into Christian theology and then deposited in political epistemology to put a face – but an impossible one – on the dream of total and complete knowledge.' For a different take on the value of framing ancient texts in light of the Anthropocene, see Formisano pp. 57–8 in the present volume.
4. Haraway 2016 and Tsing 2017 are among the most celebrated recent examples of multispecies storytelling. For both authors, 'economy' is more than an organic metaphor: their stories of multispecies sympoiesis are situated in the ecological ruins left in capitalism's wake.
5. Cf. the frequent reference that Haraway 2016 makes to Marilyn Strathern's work on 'partial connections'.
6. This aspect of the poem, signposted by its designation in the proem as a *perpetuum . . . carmen* ('continuous poem'), has preoccupied critics in search of its structural rationale for decades. See Wheeler 2000: 48–54 for discussion of some of the approaches that have been taken to accounting for the sideways (and backward) movement of its narrative transitions; and Barchiesi 2002 for the role played by inset narratives in confounding its teleological drive.
7. 'Ongoingness' is the term that Haraway 2016 coins to describe those present tense modes of adaptation and sympoiesis required of humans in order to live responsibly with other species as co-survivors of a damaged planet for a future that she refuses to foreclose. The teleological uncertainty of this narrative model offers a deliberate challenge to the apocalyptic narratives told under the sign of the Anthropocene.
8. Cf. Feeney 2007: 172–89.
9. Cf. Feeney 1999; and Gildenhard and Zissos 1999 for various accounts of these games with time in the *Metamorphoses*.

10. Van Dooren 2014: 27.
11. Rose 2012.
12. Van Dooren 2014: 27.
13. Fantham 1979: 332 draws attention to how Ovid's description of the sea storm, in particular, offers an elaborate expansion of earlier versions of the myth.
14. Hardie 2002a: 272–82.
15. *Iliad* 9.561–4 ('Cleopatra of old in their halls had her father and honoured mother called Alcyone by name, because the mother herself, having the plight of the halcyon bird of many sorrows (μήτηρ ἀλκυόνος πολυπενθέος οἶτον ἔχουσα), wept'). See also Euripides *IT* 1091–2.
16. Aelian *N.A.* 7.17: 'The ceryl (κηρύλος) and the halcyon feed side by side and live together. And when the ceryls are feeble with age, the halcyons place them on their back and carry them about upon their middle wing feathers, as they are called.' Aelian goes on to contrast the halcyons' devotion to their mates with the disregard of humans (both men and women) toward their aging spouses.
17. Aristotle *H.A.* 542b: 'Birds as a group, as has been said already, pair and breed for the most part in spring and early summer, except the halcyon. This bird breeds at the time of the winter solstice. Hence when calm weather occurs at this period, the name halcyon days is given to the seven days preceding and the seven days following the solstice . . .' Some commentators argue that Aristotle never explicitly gives us the most incredible detail of halcyon lore – the belief that the halcyon builds its nest on the sea. But see also Aristotle *H.A.* 616a on the halcyon's nest: 'It is a problem what she makes the nest of: it is thought to be mostly out of the bones of the needlefish; for she lives by eating fish.' Aristotle's suggestion here that the birds are compelled to build their nests out of fish bones seems to suggest that they are out at sea, where they would not have access to regular tree matter. If Aristotle is coy about specifying the birds' breeding location, Ovid only compounds the ambiguity: some modern translators interpret the verb *pendo* (in line 746 of *Met.* 11) to mean that the nests are made in cliffs that hang over the sea, rather than floating on the sea itself. Yet the phrasing may be deliberately ambiguous, as if Ovid were commenting on the hesitation in the ethnographic tradition about this incredible detail by describing it in words that can be taken in different ways.
18. Van Dooren 2014: 29–34 draws particular attention to the unevenness of the temporalities that intersect here: the antiquity of the albatross as a species contrasts with the very recent lifespan of the plastics with which it has become disastrously entangled.
19. Van Dooren 2014: 32.
20. Rose 2012: 128–31.
21. Cf. esp. *Met.* 11.444–5: 'And now as a hawk, friendly to none, he vents his rage on all birds. Grieving himself, he becomes a cause of grief to others' (*et nunc accipiter, nulli satis aequus, in omnes / saevit aves aliisque dolens fit causa dolendi*).
22. The story is told at *Met.* 11.749–95.
23. Haraway 2016: 8 on the current need for speculative fabulation; and 134–68 for her own foray into this genre of writing with the series of 'Camille stories'.
24. Thompson 1894: 31: 'The myth of the Halcyon days is unexplained. The above statements have no zoological significance. The kingfisher neither breeds at four months old, nor lays five eggs, nor nests in the winter season, nor on the sea. I conjecture that the story originally referred to some astronomical phenomenon, probably in connection with the Pleiades, of which constellation Alcyone is the principal star.'

25. Lucian's parody of the lore surrounding the halcyon's nest in the *True Histories* (as seen in his description of the gigantic nest floating on the sea at *VH* 2.40) reveals the scepticism with which this ethnographical lore came to be received in antiquity.
26. In his Levinasian study of the ethical responsibilities incumbent on those who witness irreparable acts of violence, such as the Shoah, Hatley (2000: 60) argues that each generation is addressed by the lives it inherits, and bears the responsibility of transmitting a memory of those lives to the next generation, 'in a spirit of commemoration and reverence as well as one of criticism and shame'. Mourning is the process that binds the past life of an individual into the historical community, the ritual that transforms an individual life into social property, and ensures that that life will transcend its own lived moment and address a broader community in the future. In this way, mourning rituals give definition to the generational differences that punctuate time for the *genos* (or, in this case, the species). Cf. Rose 2012: 130–1; and Van Dooren 2014: 9–10 for attempts to apply Hatley's ethical insights about the Shoah to contemporary extinction narratives of various species.
27. Cf. Heidegger 1962: 246–8.
28. Derrida 1993: 75–6.
29. Wolfe 2003: 146–7.
30. That surplus is hugely exaggerated in this instance by the fact that they are said to perform this annual self-sacrifice *en masse*.
31. Cf. Hopkinson 2000 *ad loc.* on the regular performance of gladiatorial games at the *Parentalia*. The origins of gladiatorial games had long been linked by ancient Romans to Roman funeral custom: Livy (*Periochae* 16.6) dates the first gladiatorial games in Rome to the funeral games held for Brutus Pera in 264 BCE by his son, Decimus Junius Brutus Scaeva.
32. The degraded status of the gladiator is commonly linked to the oath that he was said to swear upon entering his gladiator barracks, promising to endure being burned, bound, beaten and killed by the sword (to paraphrase the formulation of Petronius *Sat.* 117). It was the forfeiture of physical integrity that this oath symbolised which accounts for the gladiator's debased status. See Barton 1989: 1–4 for discussion. And see esp. Sen. *Ep.* 7.3–5 for the idea that the process of watching gladiatorial combat could turn human spectators into animals, as Seneca compares their baying for blood at the *meridianum spectaculum* ('midday games' to the behaviour of the animals in the morning beast hunts.
33. Hopkinson 2000 *ad loc.* notes the variety of manuscript readings for *more*, which include *voce* and *morte*. As Hopkinson notes, *Parentali . . . voce* would, 'refer at once to the cry of the birds as they fight, to the ritual calling of the names of the dead at the Parentalia, and to the customary cry of gladiators.' But see Hardie 2015, *ad loc.* for scepticism about the historicity of the gladiators' salute.
34. Ovid himself is one of the most widely cited sources for the idea that Graeco-Roman gods cannot weep: cf. *Fasti* 4.521–3 for his explicit statement to this effect.
35. Although, as a minor deity, like Thetis (who also mourns a dead son – and whose behaviour Aurora emulates throughout this episode), Aurora's liminal status grants her the capacity to sympathize with the human by crying.
36. Hopkinson (2000, *ad loc.*) notes the presence of this pun.
37. This view, represented by e.g. Lloyd 1968: 88–90, is based on passages in Aristotle *GA* 2.1, *De anima* 2.4, and *GC* 2.11. But see Lennox 2001b: 131–59 for important nuances to Aristotle's argument in these passages.
38. Ancient authors did harbour ideas of species extinction: see Pliny *HN* 19.15 on the extinction of the plant silphium. Parejko 2003: 927 notes that Pliny also conceives of the possible

extinction of a number of bird species, including three bird types once used in Etruscan purification rites (the *spinturnix*, *subis* and *clivia*), and the *scops* (a type of owl mentioned in the *Odyssey*). In his account of the aftermath of the Flood, Ov. *Met.* 1.436–7 notes that the earth in part (*partim*) restored the ancient forms (*figuras . . . antiquas*) of former creatures, and in part created new creatures (*nova monstra*), a phrase which would seem to imply that some pre-diluvian creatures were not restored.

39. On queer ecologies, see Haraway 2003; and Mortimer-Sandilands and Erickson 2010. On queer ecologies in Ovid, see Martelli 2020: 48–51.

CHAPTER 5
ARE TREES REALLY LIKE PEOPLE?

Emily Gowers

My title is a twist on a much-repeated opening sentence from a discussion of trees in Senecan drama.[1] 'Trees are like people' heads Robin Nisbet's short summary of the central and by now familiar bodily metaphors with which the Latin language tied trees and people together:[2]

> [Trees] have a head (vertex), a trunk (truncus), arms (bracchia). Their life moves in human rhythms, which in their case may be repeated: sap rises and falls, hair (coma) luxuriates, withers, drops off.
>
> Nisbet 1987: 202

Nisbet's analogies derive in the first place from ancient naturalists such as Columella and Pliny, who had expanded them in many directions.[3] Like humans, trees leak liquids – in their case, resin or milky sap (Pliny, *HN* 16.181). Trees have skin, blood, flesh, sinews, veins, bones and marrow (Pliny, *HN* 16.181 *cutis, sanguis, caro, nerui, uenae, ossa, medullae*). They contract arthritis; they grow obese; they marry or stay celibate; their extremities, like our hair and nails, need frequent trimming.[4] Notably, they exhibit an exaggerated version of human aging: 'In all trees, bark becomes more wrinkled in old age' (Pliny, *HN* 16.126 *omnibus in senecta rugosior*). Correspondingly, people as they age become more like trees, wrinkled and bent double.[5] Just as likely, Nisbet's list is inspired by Ovid, who had drawn on this technical terminology in the *Metamorphoses* to capture the 'rightness', or 'natural' slipperiness, of transformations from human to tree form.[6] 'Trees are like people' is, for example, the dynamic that generates his tale of Philemon and Baucis, an old couple who live in the middle of Phrygia, oldest country in the world, in the very middle of his poem.[7] After all, these wrinklies, along with the woody contents of their house and table, are already halfway to their eventual tree transformation.

These days, post-human thought and scientific research, in questioning narrow, anthropocentric criteria for agency, have made trees if anything even more like people. We know that trees, far from being lone sentinels, live in social communities, support and signal to each other through their canopies and root-systems, even feel what could be classed as desire, for feeding, reproduction and growth.[8] Philosophers like Plato, Aristotle, Cicero and Plotinus wrestled over these possibilities long ago. Could plants have souls, and were these the same as the impulses (*dunameis*) that made them feed, grow and reproduce? Did plants have a direction, a *telos*, if only to mature and replicate themselves, so achieving a kind of surrogate immortality (Aristotle, *De anima* 415a)?[9]

'Trees are like people' could also be the unspoken motto of Richard Powers's Pulitzer-winning eco-epic, *The Overstory* (Powers 2018), about nine people and their relationships with trees. The novel even takes the shape of a tree, tracing the roots of its protagonists, building a trunk at its core and ending with seeds for the future. Centuries before, Ovid's *Metamorphoses* had an equivalent 'overstory' told by the steady presence of its trees, one often overlooked in favour of the more dramatic panorama of its human events. Ovid casts his quizzical eyes over these silent witnesses, seeking traces of humanoid physicality and sentience in an imagined world where trees preserve human consciousness inside their rugged bark. His many memorable descriptions of humans morphing into trees have inspired modern poets like Ezra Pound, John Ashbery and Philip Larkin:

> The tree has entered my hands,
> The sap has ascended my arms,
> The tree has grown in my breast –
> Downward,
> The branches grow out of me, like arms.
>
> Ezra Pound, '*A Girl*'

Ovid's transformed trees share their space with humans who have escaped transformation, reproaching them with a different pace of life, providing a backdrop that enables worldly cohabitants to reflect on their own existence. Story-teller Orpheus is surrounded by an audience of transformed laurels, oaks and lindens, while Narcissus stretches out his human arms (*bracchia*) to meet the tree-arms (also *bracchia*) of the silent watchers that stand around him:

> ad circumstantes tendens sua bracchia siluas
> *stretching out his hands to the by-standing woods.*[10]

As Owain Jones and Paul Cloke claim: 'There is nothing more stubbornly "there-in-the-world" than trees.'[11] Or, as John Ashbery puts it, in *Some Trees*: 'their merely being there | Means something . . .'

Yet one thing seems clearer to me now: the meaning of trees in Ovid has just as much to do with how they *aren't* like people as with how they are. Their rigid bark stands in the way, for a start (though humans grow into that with sun and aging); so does their rooted immobility. Bernini's sculpture of Apollo and Daphne invites its viewers to rotate – curiously, and even cruelly – around its spiralling shape, at the very moment, frozen from Ovid's narrative, when Daphne's swift feet are glued to the ground by tenacious roots (1.551 *pes modo tam uelox pigris radicibus haeret*). Trees have no language or at least they don't make intelligible noises (when Larkin wrote, in 'Trees', 'The trees are coming into leaf | Like something almost being said', 'almost' is the resonant word). Trees have seen more but have no eyes, except the stumps of their lost branches or the holes where a graft fits in (a process Latin calls *inoculatio*, inserting via the 'eyes').[12] In Rome, if a tree refused to stand still or made a noise, it was classified as a sinister portent. Pliny describes lakeside oak trees that when uprooted dredge up vast islands of soil and float along standing upright, terrifying ships with the 'wide rigging' of their branches' (*ingentium ramorum armamentis*). He marvels equally at the vast Hercynian oak forest, untouched by the ages and coeval with the world:

> In the same northern region is the vast expanse of the Hercynian oak forest, which surpasses all other marvels by its almost immortal destiny. To omit other facts that would lack credence, it is well known that the collision of the roots encountering each other raises up hillocks of earth, or, where the ground has not kept up to them, their arches in their struggle with one another rise as high as the branches, and curve over in the shape of open gateways, so as to afford a passage to squadrons of cavalry.[13]

Pliny also writes about the crackling noise (*strepitus*) to be heard when the mulberry tree's buds sprout in a single night (16.102) and notes that door hinges made of olive wood are prone to put out shoots, like a growing plant (16.230). He even attributes emotions to plants: distaste, obstinacy, stupidity . . .

Ovid's primeval forests are no less static. When Cadmus, founder of Thebes, invades a wood untouched by any axe, he must disturb the thickets standing in his way: *obstantes proturbat pectore siluas* ('he ruffles the oncoming woods with his chest').[14] The tree on which his enemy, the incumbent serpent, sprawls, lashing its tail, takes on not only the animal's sagging shape but also its humanoid voice: *pondere serpentis curuata est arbor et ima | parte flagellari gemuit sua robora caudae* ('the tree was bent by the serpent's weight and groaned when its timber was lashed by the tip of its tail').[15] The Calydonian wood in *Met.* 8 is a virtual obstacle course of animated trees, which trip hunters up furtively with their roots or knock them senseless with stray branches. Ovid, like Pliny, was all too aware of what Jones and Cloke call the 'unruly agency' of trees, the hidden but often explosive mystery of their growth.[16]

But perhaps the most important factor that makes trees different from humans has to do with time. Most obviously, trees live longer and grow more slowly. The narrative of *The Overstory*, for example, alternates between human time and tree time, marking the difference between our lifespan and what Powers calls 'the speed of wood', that is, slow and imperceptible. Today, the speed of wood is getting faster. In California, the giant sequoias of the Sierra Nevada have long been a barometer for climate change. These most resilient of trees, sometimes as old as three thousand years, have a cult history dating from the 1850s, when national sentiments were aroused by the unnecessary felling of a single 'mammoth' tree. Today's record-breaking specimens – some of them with curiously Ovidian names (Daedalus, Icarus, Hyperion, even Smashing Titans) – have survived environmental stresses relatively well and can even claim to provide some resistance, given their potential for cloning. Nonetheless, their lifecycles are speeding up and the trees are dying sooner.

The ancients were equally aware of the time differential between humans and trees; in fact, they tended to exaggerate it. Looking to the remote regions of the world and their impenetrable forests, Pliny allows that some trees have 'an immeasurable lifespan' (*HN*

16.234 *uita ... immensa*). He recalls ancient trees in Rome, older than the city, and mythical trees shown to tourists in Greece and Phrygia, like the olive to which Argus tethered Io or the tree on which Marsyas was hanged. On the other hand, the annual, seasonal cycle of trees' growth made them a speeded up, self-replenishing equivalent to human life. Recall Nisbet's second sentence: *Their life moves in human rhythms, which in their case may be repeated*. Homer (*Iliad* 6.146–9) had looked to deciduous leaves, the most impermanent part of trees, as an analogue for human life.[17] Mark Payne has recently written about John Ashbery's 'Some Trees' and the counterintuitive notion of shallow time that those trees suggest.[18]

This chapter explores how Ovid uses trees in *Metamorphoses* to map chronology in a way that potentially works against the beat of human time and other kinds of time, historical and metamorphic.[19] It asks what kinds of transformative change the poet allows a tree population that has existed since Day Two of the universe, before human life began, when the creator clothed the new land with woods and the woods with leaves (*Met.* 1.44 *fronde tegi siluas*). How, in particular, do his trees fit with Thomas Cole's assessment of *Metamorphoses* as universal history in *Ovidius Mythistoricus* (Cole 2008)? Cole detects a steady process of aging among the human dramatis personae as the poem goes on: young characters fill the first third of the poem (except for Tiresias, Ino and Coronis, everyone in Books 1–5 is a child or adolescent); in the middle third, Books 6–11, most humans are married, or old enough to be (heroes at Troy have sons and daughters; women are mothers, not wives); and finally in Books 12–15 come Nestor, Caieta, the thousand-year-old Sibyl, King Numa and Pythagoras.[20] Looked at more systematically, this scheme turns out to be an over-simplification, or at any rate a rule with many exceptions. Even so, it remains a useful thought-experiment for reflecting on the possibility of ontogenic and phylogenic coincidence in *Metamorphoses* – that is, the simultaneous progress of aging individual bodies and the aging of the human race as a whole.

Cole argues that ancient thinkers had their own version of biogenetic law, as formulated in 1886 by Ernst Haeckel (who invented the term Oekologie, ecology) and revived by Stephen Jay Gould in 1977. According to this scheme, 'the long-term evolution of a collective entity . . . passed through stages parallel to those subsequently recapitulated in shorter periods by its individual components.'[21] Haeckel had applied this 'law' to animal embryos, whose development and birth (ontogeny), he alleged, was patterned on the evolutionary history (phylogeny) of the species as a whole.[22] Such a scheme had been anticipated millennia before, when Roman historian Florus patterned the ages of Rome's history on the ages of man: the era of kings was its childhood, the early Republic its fiery youth, the reign of Augustus its maturity and the subsequent imperial period its decline – with the exception, predictably, of the vigorous current ruler, Trajan (1 *praef.* 4–8). What is more, Ovid himself gives us a variant on this scheme in *Met.* 15, where the sage Pythagoras describes human life as patterned on the calendar year:

> nam tener et lactens puerique simillimus aeuo uere
> nouo est; tunc herba recens et roboris expers
> turget et insolida est et spe delectat agrestes.

omnia tunc florent, florumque coloribus almus
ludit ager, neque adhuc uirtus in frondibus ulla est.
transit in aestatem post uer robustior annus
fitque ualens iuuenis; neque enim robustior aetas
ulla nec uberior nec quae magis ardeat ulla est.
excipit autumnus, posito feruore iuuentae
maturus mitisque inter iuuenemque senemque
temperie medius, sparsus quoque tempora canis.
inde senilis hiems tremulo uenit horrida passu,
aut spoliata suos aut quos habet alba capillos.

Met. 15.201–13

For spring, in its new life, is tender and milky, like a child: then the shoots are fresh and still unformed, delicate, without substance and delighting the farmer's hopes. Then everything blossoms, the bountiful land is a riot of coloured flowers, but the leaves are still not strong. From spring the year, grown stronger, moves to summer, and becomes a powerful man: no season is sturdier or more abundant, than this, or hotter. Autumn takes over, when the ardour of youth is gone, ripe and mellow, between youth and age, a scattering of grey on its brow. Then trembling winter, its hair despoiled, or, what it has, turned white.

Not only is human aging mapped onto the annual cycle here, but the adjectives used are those of vegetable growth: *recens*, *robustior*, *maturus* and *mitis* suggest bud, timber and fruit; Pythagoras entwines the lifecycle of plants with that of both the human body and the seasons.[23] His summary suggests some retrospective questions to put to Ovid's text. How neatly can the phylogenic lifecycle of trees in *Metamorphoses* be superimposed on its larger ontogenic narratives? Can we draw conclusions from the line-up of arboreal stories about the ultimate telos of the poem? And if so, does that telos consist in maturation, decay, renewal or even self-combustion?

Cole also argues, briefly, that the poem's ecosystem changes in stages, too. By the end of Book 1, humans have moved from virgin forest to pastures, then in Book 2 to city-building.[24] But there he stops, and it remains unclear what if any trajectory can be extrapolated for the rest of the poem. True, Rome is the endpoint, with Augustus and the domestic cult centred on the Palatine, symbolized by a confined space, the orchard precinct (*pomarium*) of Pomona. But there are plenty of reversions to wild space before then: Pentheus on Mount Cithaeron, Hermaphroditus in the wilds of Caria, and so on. In between, signs that Ovid's trees map a generic progress are hard to find. Indeed, an opposite trend seems to emerge instead: trees acting as a counterpoint to human evolution. This is a hypothesis worth testing across the full span of the poem. For it is not just trees' cyclical rhythms that complicate issues of simple teleology. It is also the coexistence of tree and human bodies, whether in the form of a nymph who already belongs to a tree or a human suddenly turned into one.[25] Ovidian trees superimpose two layers of time, encasing a younger, or younger-looking body inside an older, or older-looking one. These layered arboreal anatomies respond to questions that have haunted religious thinkers almost from the beginning. Do plants have individual consciousnesses? Do they contain independent spirits with their own souls, locked inside the bark?[26]

This debate was long focused on tree nymphs, dryads or hamadryads. (Alternatively, these imaginative and intellectual questions were the reason such creatures were invented.) Was a hamadryad the same as a tree, a spirit specific to one tree, or a god present in all trees of the same type?[27] While Greek *hama* suggests identity or simultaneity, the assumption that nymphs were identical with their trees has lingered thanks to a few ancient commentators.[28] In practice, dryads and hamadryads were interchangeable, and even hamadryads are found acting autonomously. Ovid's Erysicthon contributes to the debate when he cuts down a tree with a dryad in it:

> non dilecta deae solum, sed et ipsa licebit
> sit dea, iam tanget frondente cacumine terram.'
>
> *Met.* 8.755–6

> *Even if this is the goddess herself, and not just what the goddess loves, now the tree will touch the earth with its leafy crown.*

Time plays its part here, too. Not only were trees and nymphs potentially coexistent: often, both were thought to be very, very old. Hesiod (fr. 304) had claimed that nymphs live for exactly 9,720 human generations; Pindar (fr. 165) alleged that they were coeval with trees. Callimachus ponders these questions of coexistence and equal age:

> *My goddesses, you Muses, say whether the oaks came into being at the same time as the Nymphs. The nymphs rejoice when the rain makes the oaks grow; and again the Nymphs weep when there are no longer leaves upon the oaks.*
>
> Callim. *Hymn to Delos* 4.82–5

But the most extended of all discussions about identity, age and immortality in relation to trees and nymphs can be found in the *Homeric Hymn to Aphrodite*, where the goddess uses her one-night stand with Anchises, father of Aeneas, to reflect on the unviability of all divine–mortal unions. Negotiating with Zeus to keep her human lover, she recalls the cautionary example of Dawn (Eos), who asked for immortality for Tithonus but forgot to ask for eternal youth as well, so was forced to hide her bent, decrepit husband in her bedroom. But Aphrodite herself fails to ask for immortality *and* eternal youth for Anchises. It is as though deep-down she accepts the short-lived nature of their encounter in a hut at the top of a mountain (something her son will go on to repeat in his alfresco 'marriage' to Virgil's Dido). In the passage that follows, she describes the nymphs to whom she plans to farm out her infant son:

> *As for [Aeneas], once he sees the sunlight, he will be nursed by the deep-bosomed, mountain- couching nymphs who dwell on this great and holy mountain, who belong*

with neither mortals nor gods. They have long lives, and eat divine food, and tread the fair dance with the immortals; Sileni and the keen-sighted Argus-slayer join in love with them in the recesses of lovely caves. As they are born, fir trees or tall oaks come forth on the earth that feeds mankind: they stand fine and healthy, towering in the high mountains, and people call them precincts of the gods, and mortals do not cut them with the axe. But when their fated death is at hand, first the fair trees wither where they stand, their bark decays round and about them, their branches fall off, and simultaneously the nymphs' souls leave the sunlight.

Hymn to Aphrodite 256–72

The main reason Aphrodite focuses on nymphs and their trees is that their shared lives span the mortal–immortal divide.[29] But this is not the only phenomenon she considers. Whatever their ontological relationship, nymphs and trees also conflate different *times* in one organism (Aphrodite refers to nymphs as 'clothed in wood', 285).[30] After all, their coexistence is surprising, given that trees are wrinkled and rooted and nymphs perennially youthful and perpetually mobile, dancing outside and sleeping with sileni. When Aphrodite describes the joint death of nymphs and trees, she homes in on the trees using a hapax, *amphiperiphthunei*, a compound of *pithuno*, 'to wither, waste away'. In Homer, the verb is interchangeable with *phthio/phthino* and often joined to one or other prefix (*amphi* 'around' and *peri* 'about') that suggests the sheer physical roundedness and layeredness of a tree. As it dies, a tree's bark will 'wither around and about' it, not just on its exterior but all around its circumference, leaving the nymph inside exposed and vulnerable. Ovid inverts this idea, again in his Erysicthon episode, where he has dryads encircle a great oak tree with their arms (*Met.* 8.748–9 'the girth of its timber measured fifteen cubits').

This startling image of 'withering around and about' itself inverts a metaphor used twice earlier when Aphrodite comments on Dawn's careless request:

To be sure, I would not choose for you to dwell in that form among the immortals, to be immortal and live for all days. But if, being such as you are in appearance and build, you might live, and you might be called my husband, then grief would not enfold my clever wits. But now quickly baneful old age will enfold you, pitiless – and this stands beside men ever after, destructive, debilitating, which even the gods hate forever.

Hymn to Aphrodite 239–46

If Anchises could become immortal and eternally young, grief would not 'enfold' or 'clothe' (*amphikaluptoi*) her mind. As things stand, old age 'will enfold him' (*amphikalupsei*, the future tense of the same verb). As in Homer, the bark of trees functions as a metaphor for sorrow, death and aging, whether flaking off, exposing a raw interior or enveloping in a wrinkly crust. The nymphs inside will eventually die, but a human female is destined for withering both ways. Trapped inside a tree, she won't marry happily (like Dawn, trapped with her wrinkly husband); living outside, she will be clothed with sorrow and old age, dry up and die soon anyway (Greek *nymphe* 'nymph' also means a bride or nubile young woman; *parthenos* is the word used here, perhaps to avoid confusion). Aphrodite's thoughts, I believe, offer a useful model for the *Metamorphoses*: a way of thinking about trees as simultaneous conflations of different timeframes, physicalities and stages of life, even as they age for far longer than humans do.

Let us turn now to the tree episodes of *Metamorphoses*. It would be impossible to do justice to them all. Ovid provides so many extraordinarily graphic variations on the theme of a human being, usually a woman, helplessly swallowed up by bark and losing mobility and voice that such episodes seem to swim into each other with a kind of déjà vu inevitability.[31] It helps to remember how much our attitude to each metamorphosis is shaped by how we read the poem as a whole. As Andrew Feldherr has shown, we can choose between two very different perspectives on any metamorphic event. From one viewpoint, it is a tragedy, in which the human is engulfed and deprived of their former life; from another, it is closer to comedy, as one story yields to the next in a broader narrative flow.[32] Alternatively put, the choice is between a human perspective and a divine one. In the case of tree metamorphoses, the speed of the transformation is often startlingly at odds with tree time, as if human life, relatively speaking, were a split second or the twinkling of an eye, absorbed into the longue durée of arboreal life. Can we conclude that the trees, too, just go cruelly on? After all, one or two or more are added to their number each time; after each human loss, the earth is reforested.

With Thomas Cole's template of progressive aging in mind, it is striking that the first tree metamorphosis, that of the nymph Daphne, is into not a blossoming or deciduous tree but an evergreen one (she is after all a nymph, not a human female). Daphne will endure until the very end of the poem: she remains a perpetual virgin in the audience of trees that gather round Orpheus in Book 10 (92 *innuba laurus*), and in Book 15 it is laurel that hides Cipus' horns (591 *pacali . . . lauro*) and trembles sympathetically at Delphi (634). Daphne's eternal youth as nymph, nubile girl or arrested bride-to-be is marked by evergreen-ness, not blossom. Virginity is cast in arboreal terms as a lack of transition to the next stage, in which she herself seems to be complicit – witness her ambiguous nod (*Met.* 1.567 *adnuit*). It is also a condition that carefully binds one end of the poem to the other.[33]

Something similar happens with the next tree-victims, Phaethon's sisters, the Heliades, once Earth has turned to toast after their brother's chariot accident:

> non satis est: truncis auellere corpora temptat
> et teneros manibus ramos abrumpit; at inde
> sanguineae manant tamquam de uulnere guttae.
> 'parce, precor, mater' quaecumque est saucia clamat,
> 'parce, precor; nostrum laceratur in arbore corpus.
> iamque uale' – cortex in uerba nouissima uenit.
> inde fluunt lacrimae stillataque sole rigescunt de
> ramis electra nouis, quae lucidus amnis
> excipit et nuribus mittit spectanda Latinis.

Met. 2.358–66

That is not enough: she tries to tug their bodies from the trunks and breaks the growing branches from their hands, but bloody drops seep from there, as if from a wound. 'Save me, mother, I beseech you,' her wounded daughters shout, 'Save me, I beg you: my body is torn apart inside a tree: but now, farewell' – bark covered the end of her words. Tears flowed there, and amber distilled from the new branches grew hard in the sun, which the shining river received and sent off to be admired on Roman daughters-in-law.

Following their metamorphosis into poplars, these girls, like so many others, are deprived of marriage and reproduction (there is a pointed focus on *uterus*, womb, and *pectora*, breasts). Yet they become a tappable source of the resin that will, in Ovid's time, deck Roman brides (*nuribus Latinis*, 'Latin daughters-in-law', indicates the further rite of passage that is marriage). Amber drips from their trunks, like tears or blood from a wound, before hardening and crystallizing them in a prenuptial state (consistent with Cole's schema), even as their product enables later women to enter a new life-stage. More interestingly, from a global point of view they are the prelude to a larger project of eco-restoration. Requiring a sylvan backdrop to seduce Callisto, Jupiter restores the dried-out leaves of Arcadia and revives its damaged woods, in a speeded-up, frivolous, divinely convenient version of natural regeneration:

dat terrae gramina, frondes arboribus,
laesasque iubet reuirescere siluas.
. . .
cum subit illa nemus quod nulla ceciderat aetas

Met. 2.407–8.418

[Jupiter] gives grass to the earth, leaves to the trees, he bids the damaged woods grow green again . . . when she came across a grove which had been uncut through the years . . .

So it is that Callisto wanders into a grove that has always been there, untouched. Two ways of looking at change are contained in consecutive stories: cruel enclosure, on the one hand; harvesting, commerce and reforesting, on the other.

Book 3 ends with the image of a lone tree in an otherwise depleted landscape. Pentheus, king of Thebes, is dismembered by his female relatives. Hunted down in a natural amphitheatre 'clear of all trees' (*purus ab arboribus*), he becomes a 'tree' himself, in a memorable simile which sweeps up and rationalizes stray arboreal metaphors:

non habet infelix quae matri bracchia tendat,
trunca sed ostendens dereptis uulnera membris
'adspice, mater!' ait . . . .

Met. 3.723–5

The wretched man had no arms to stretch out to his mother, but showing the wounded stumps from lopped off limbs, said, 'Look, mother!'.

. . .
non citius frondes autumni frigore tactas
iamque male haerentes alta rapit arbore uentus,
quam sunt membra uiri manibus derepta nefandis.

Met. 3.729–31

No faster does the wind sweep away leaves touched by autumn chill, still clinging to the treetop, than the man's limbs were stripped by those wicked hands.

First, the king stretches out his *bracchia* (arms/branches), then he waves his *trunca uulnera* (mutilating wounds, or holes for grafting) – before Ovid reins in his imaginative leap (turning metaphor to simile), concluding only that Pentheus is *like* a deciduous tree, limbs dropping like autumn leaves (recalling the tree lookout on Mount Cithaeron from which the same character watches the revels in Euripides' *Bacchae*, conflated with his opposite number, his cousin Dionysus, a tree god).[34] Pausanias (2.2.7) tells how the inhabitants of Corinth were instructed by the Pythian priestess to find the tree that Pentheus hid in and turn it into wooden statues of Dionysus. Ovid similarly superimposes Pentheus on Bacchus when he equates the king of a failing city with a tree in decline, like Virgil's truncated Priam, or a tree in defeat, like another despiser of the gods, Virgil's Mezentius, whose armour clothes a tree- like trophy. This is, in short, a moment when human, city and vegetable life seem to be in alignment.

Book 4 offers another story of failed human maturation: Pyramus and Thisbe. The bloody death of one unfortunate Babylonian lover causes the white mulberry to sprout blood-red berries, giving a lively retrospective aetiology for Virgil's dead metaphor in the *Eclogues* (*Ecl.* 5.22 *sanguineis . . . moris*, blood-coloured mulberries):

arborei fetus adspergine caedis in atram
uertuntur faciem

Met. 4.125–6

With the spattering of blood, the tree's fruit changes to a black colour

Here, trees and people behave very differently. The couple's arrested development, their inability, despite increased proximity, to touch each other except in death, is countered by the tree's onward evolution, as it reveals the newly black appearance of its fruit. There is sadness, now, behind Ovid's statement of fact:

nam color in pomo est, ubi permaturuit, ater

For the colour of the fruit, when it has ripened, is black

Met. 4.165

By contrast with the unlucky pair who did not mature, the tree at least went on producing (Thisbe commands it to 'always bear fruit', 4.161 *semper habe fetus*), providing a merciful canopy for the couple's ashes. Shortly afterwards, when Salmacis and Hermaphroditus violently coalesce into one intersex being, Ovid compares them to a graft inserted into a tree so that the two grow (up) together (4.376 *pariter adolescere*) – a more positive image of simultaneous and equal maturing than the one that evolves in the main story, where Hermaphroditus is seemingly reduced to an inadequate version of his former self (*semimas, semiuir*).[35]

Past Book 7 (where the Myrmidons evolve from a tree's parasites, little black ants) and Ovid takes stock (no pun intended) at the woody heart of the poem. Trees crowd in now – not just the hunters in the Calydonian wood but also Philemon and Baucis and Erysicthon's sacrilegious oak-cutting.[36] This central book layers primitivism onto renewal, allowing stories of destruction and regeneration (a second flood, another Mr and Mrs Noah, following Deucalion and Pyrrha in *Met.* 1) to be read simultaneously. Not only do Philemon and Baucis's transformed house and tree bodies link past and present (Phrygian reeds to Augustan marble), but the pile-up of domestic timber that surrounds the two geriatrics in their woody hut is sacrificed in the form of Meleager's metonymic log, then subsumed into Erysicthon's gargantuan bonfire of *materia* (material, or, more literally, timber). The culminating simile compares insatiable hunger to an all-consuming forest fire:[37]

> utque rapax ignis non umquam alimenta recusat
> innumerasque trabes cremat et quo copia maior
> est data, plura petit turbaque uoracior ipsa est.
>
> *Met.* 8.837–9

> *Just as a rapacious fire never refuses fuel and burns unquantifiable amounts of timber, and the more supply is given to it, the more it demands and it becomes too greedy for the piles it is given.*

From these ashes emerges the tree-metamorphosis of Dryope (*Met.* 9). Here, at last, is a story to support Cole's schema.[38] This time, it is a mature woman, a mother, whose transformation yields an aetiology for a child's weaning. Dryope makes the mistake of plucking flowers for her baby Amphissus from a sacred lotus tree, as her sister recalls:

> haeserunt radice pedes; conuellere pugnat
> nec quidquam nisi summa mouet. subcrescit ab imo
> totaque paulatim lentus premit inguina cortex.
> ut uidit, conata manu laniare capillos
> fronde manum impleuit: frondes caput omne tenebant.
> 'At puer Amphissos (namque hoc auus Eurytus illi
> addiderat nomen) materna rigescere sentit
> ubera, nec sequitur ducentem lacteus umor.

spectatrix aderam fati crudelis opemque
non poteram tibi ferre, soror; quantumque ualebam,
crescentem truncum ramosque amplexa morabar
et, fateor, uolui sub eodem cortice condi.

Met. 9.351–62

. . .her feet stuck with roots. She fought to tear them away: nothing moved but her upper body. Slowly, thick bark grew upwards from her feet, hiding all her groin. When she saw this, and tried to tug at her hair with her hands, her hands filled with leaves: leaves covered her whole head. But the child Amphissos (so his grandfather, Eurytus, King of Oechalia, had named him) felt his mother's breast harden, and the milky liquid failed when he sucked. I was there, an observer of your cruel fate, sister, and could give you no help. With all my strength, I held back the growing trunk and branches with my embrace, and, I confess, I wished I could be buried under the same bark.

As Dryope is translated into hard tree-matter, her milk dries up. Pliny (*HN* 16.181) regards the liquid juice in the body of trees – milky, slimy, gummy, sticky or watery – as their 'blood'. But Ovid puts the two liquids in counterpoint: the tree nymph bleeds, while Dryope's own breasts harden (10.357 *rigescere*) and their milky liquid (358 *lacteus umor*) fails when the child tries to nurse. The story recalls certain victims in the *Aeneid*: Dryops, whose voice is suffocated by a wooden spear (*Aen.* 10.346–9), and Tarquitus, son of Dryope and wood-dwelling Faunus, beheaded and left with a still-warm trunk (*Aen.* 10.555 *truncum tepentem*). Once again, the event is a tragedy for Dryope and her bereaved family. Yet in the larger scheme of things it results in reforestation – either by cloning, if the manuscript reading *loton* 'lotus-tree' is accepted, or diversification, if an oak or a poplar is preferred.

Book 10 yields another thick cluster of trees. Not just the arboreal audience for Orpheus' string of stories, transformed humans charmed into regrouping to answer his need for story-telling shade (10.88–105). Among his subjects are Cyparissus, Apollo's beloved turned into a cypress, and Myrrha, punished for incest with her father Cinyras by being transformed into a myrrh tree. At first, Myrrha resembles a traditional victim, a non-bride duly engulfed into arboreal form (a proleptic simile, 10.372–4, even compares her to an axed tree as she hesitates, not knowing which way to fall). Her tree works as a particularly good image of the conflation of different times I mentioned earlier. A young woman with still-warm branches and resinous tears is trapped inside a hard outer bark:[39]

nam crura loquentis
terra superuenit, ruptosque obliqua per ungues
porrigitur radix, longi firmamina trunci,
ossaque robur agunt, mediaque manente medulla
sanguis it in sucos, in magnos bracchia ramos,
in paruos digiti, duratur cortice pellis.

Met. 10.489–94

For as she spoke, the earth overcame her, and roots, breaking from her toes spread sideways, supporting a long trunk, and her bones grew tough, and in the middle of her remaining marrow her blood became sap, her arms became long branches, her fingers twigs, her skin solid bark.

The tree encapsulates the generationally twisted union of father and daughter: a soft young female encased inside an older being, just as the father 'received his own offspring/ guts (*uiscera*) in his bed' (10.465). But the generations coalesce further when Myrrha's child, Adonis, bursts forth. The description of her bark splitting as it opens to let the baby out is particularly savage:

> At male conceptus sub robore creuerat infans
> quaerebatque uiam qua se genetrice relicta
> exsereret; media grauidus tumet arbore uenter. Tendit
> onus matrem, neque habent sua uerba dolores, nec
> Lucina potest parientis uoce uocari.
> nitenti tamen est similis curuataque crebros
> dat gemitus arbor lacrimisque cadentibus umet.
> constitit ad ramos mitis Lucina dolentes
> admouitque manus et uerba puerpera dixit;
> arbor agit rimas et fissa cortice uiuum
> reddit onus, uagitque puer . . .'.
>
> *Met.* 10.503–13

But the ill-conceived infant had grown deep in her fibre and sought a route by which it might leave and exit from its mother. Her womb swelled heavily in the heart of the tree. The burden stretched the mother; and her pain had no words, nor could Lucina be summoned by the labouring mother's voice. The tree was bent like someone in labour, and let out constant groaning and grew wet with falling tears. Gentle Lucina stood by the suffering branches, applied her hands and spoke consoling words: the tree sprang cracks and let out its living burden through its split trunk, and a boy wailed . . .

As usual, Ovid is interested in the correspondence of tree and body parts and the symptoms of pain that assimilate tree to human mother – creaking, contortion, splitting, resinous tears.[40] But there is one obvious mismatch. Trees do not usually produce fruit through their trunks: their reproduction works upside-down in human terms. Varro, for example, observed that the human foetus, growing normally in the womb, is like a tree, head down, feet upmost.[41] Plato in the *Timaeus* had compared man to a tree whose roots are not in the earth but in the heavens (90a) and Aristotle (*De partibus animalium* 686B) noted, conversely: 'In plants the roots have the character and value of mouth and head'.[42] From this came the idea of man as *arbor inversa* which lasted into the Renaissance and was used for comic or carnivalesque effect by Rabelais and Swift, who in his *Meditation upon a Broomstick*, called man a 'withered Bundle of Twigs' tied to a 'sapless Trunk', 'a topsy-turvy Creature . . . his Head where his Heels should be, groveling on the Earth'.[43] In the ancient world, any tree that gave birth via its trunk was sinister. Pliny (*HN* 16.199) speaks of a wild olive tree in Megara on which warriors had hung weapons, around which bark grew. When an oracle predicted that the city would be destroyed when a tree gave birth to arms, they cut down the tree and found the weapons inside it.

One expression for a tree's fruit, *arboreus fetus*, recurs several times in the poem. We have already seen it describing Pyramus and Thisbe's mulberries. It will also describe the fruit Pomona tends (14.625) and add to the bounty of the Golden Age (15.97). But its appearance in Book 10 comes with an added twist. When Venus relates the story of invincible athlete Atalanta, the crucial apple dropped by Hippomenes to distract his beloved from the race is chosen from three *fetibus arboreis* (10.665, 'arboreal fruit'). This would not be particularly interesting (nor has it interested commentators) – except that Venus is telling the story to her short-lived lover, the very same Adonis who was himself the 'strange fruit' of a portentous tree (before reverting to an ephemeral blossom, the anemone, in a misdirection of normal vegetable growth). Overall, the Myrrha episode reworks the Greek opposition between the male ephemeral festival, Adonia, and the female harvest festival, Thesmophoria, Myrrha's mother's absence at the latter creating the need for a younger woman to fill her husband's bed (*Met.* 10.431–5).[44]

By this point in the poem, a pattern of ebb and flow, rather than straightforward progress, seems to be emerging. When Orpheus dies, torn apart by Thracian Maenads, his tree-listeners (unless they are evergreen) shed their foliage (11.47 *comas*), while their incumbent nymphs tear out their equivalent hair (11.49 *capillos*). It is fitting that the trees' activity at the end of these entropic, introspective tales is deciduous, like Pentheus' stripped tree at the end of Book 3. Almost immediately, however, arboreal replenishment arrives from an unexpected source. The Maenads, hyper-mobile females, turn into oak trees, rooted like birds in lime. Other forests follow. The reeds that betray King Midas' ass's ears are described as a dense grove (11.190 *creber . . . lucus*). Sitting comfortably for the contest of Midas and Apollo, Mount Tmolus becomes a charming anthropomorphic landscape, clearing woody growth from his bushy ears with their dangling acorns (11.157–9). When transgender warrior Caeneus is killed by the centaurs, an entire mountainside suffocates him under a mass of wood, in a process that strongly resembles a female tree transformation.[45] Polyphemus even woos Galatea by telling her that every tree is 'at her service' (13.820 *omnis tibi seruiet arbor*), a motto that might be applied to the poem as a whole.

In Book 14, however, something more positive occurs, in an episode that David Littlefield long ago read as a miniature of the settlement of the Trojan dynasty in Italy.[46] Pomona, a hamadryad who definitively lives *outside* her trees, tends an ordered orchard (*pomarium*) and is fanatical about raising prize fruit (14.625 *nec fuit arborei studiosior altera fetus*), not producing her own offspring. The penultimate section of the poem has already been flooded by images of ripeness, to which her marriage to Vertumnus, god of the changing year, is just the pendant. Aeneas is ready for heaven (584 *tempestiuus . . . caelo*), Ascanius is growing into a stable future (583 *bene fundatis opibus crescentis Iuli*), even Jupiter is *mitissimus* (587, most mild or most ripe) – so that Vertumnus' rhetorical techiques of mollifying and directing his beloved complement both Pomona's arboricultural work of training and ripening and the larger political picture of consolidated settlement (593–609).[47] Against Pomona's grafting techniques (630–1 *fisso modo cortice lignum | inserit et sucos alieno praestat alumno*, 'now she grafts wood into a split trunk and provides sap for an extraneous nursling'), Aeneas (a 'vigorous sprig', *thaleron gonon*, at *Homeric Hymn to Aphrodite* 104) appears as another extraneous nursling, grafted onto native power in Italy.[48] A potential rape scene is diverted into an Ovidian triumph of amatory persuasion over barely escaped violence, in a speech filled with convincing parallels from the lives of trees: if a bachelor elm tree did not support a vine, it would boast of nothing but its leaves; if a spinster vine had no elm, it would not be laden with fruit (663–6). Even the final union is framed ambivalently as a kind of violent grafting (771 *mutua uulnera sensit*, 'she felt a mutual wound').[49] With the exception of one wild card, the oleaster or wild olive whose bitter berries conserve the caustic flavour of subversive Roman humour (14.525–7), Ovid's trees are coming home, like the gods, into the human, urban fold.

Even so, the Pomona-Vertumnus episode is a layered tissue of different times. To seduce Pomona, Vertumnus turns himself into that most tree-like of beings, an old woman; as Ovid says of Jupiter and the Cercopian monkeys earlier in the book: 14.96 *rugis perarauit anilibus ora*, 'he ploughed [the monkeys'] faces with an old woman's wrinkles'. As a god, Vertumnus can reverse the process: 'he was restored to youth and laid down his old woman's accoutrements' (14.766–7). This underscores the difference between human and divine capability, rosy youth and wrinkly trees. But it is also a graphic demonstration of the god's ability to speed up, then renew the seasons of human life.[50]

The last tree in *Metamorphoses*, however, is a dead one which sprouts miraculously into new life. Rome's early hero Romulus sees his spear, grown from a native Palatine plant, suddenly put out leaves and take root, becoming a sturdy willow:

> utue Palatinis haerentem collibus olim
> cum subito uidit frondescere Romulus hastam,
> quae radice noua, non ferro stabat adacto
> et iam non telum, sed lenti uiminis arbor
> non exspectatas dabat admirantibus umbras
>
> *Met.* 15.560–4

> *As when Romulus suddenly saw his spear which was once rooted in the Palatine hill start to sprout leaves and stand with a new root, not with its point driven in – no longer a weapon but a tree of tough willow, offering unexpected shade to those who wondered at it.*

It is not hard to read the symbolism here. Roman historiographical writing records many trees which sprout new leaves at times of political change, especially changes in dynasties, a motif ironized by Tacitus, noting that Romulus' Ruminal fig-tree revived in the darkest days of Nero's reign.[51] The spear miraculously fulfils two well-known Homeric and Virgilian *adynata*: Achilles and King Latinus swear by sceptres that will never bear leaves again (*Il.* 1.234–9; *Aen.* 12.206–11). Virgil is particularly precise about Latinus' sceptre's bronze casing, which does not just strip the leaves, as in Homer, but traps them: what was once a tree (*olim arbos*) is now enclosed by artistry (210–11 *artificis manus aere decoro | inclusit*). But Ovid's final rootstock defies expectation and continues to provide shade and spawn wonder.

Here end the trees in Metamorphoses. There is nothing so neat as the olive-themed border that rounds off Minerva's tapestry (6.101–2 *circuit extremas oleis pacalibus oras | (is modus est) operisque sua facit arbore finem*, 'She bordered the outer edges with peace-bringing olives – this was the final touch – and rounded off her work with her own tree'). And yet the adjective *perennis*, applied to Ovid's flight to heaven at the end of the poem, recalls earlier images that bind poetic and tree immortality, not just Daphne's laurel but also Lucretius' 'perennial *Ennius*', with his garland of the muses (*DRN* 1.118), and 'always flourishing Homer' (*DRN* 1.124), not to mention the Delphic laurel that binds up Horace's third book of Odes (*C.* 3.30.15–16).

We can look back more consciously now at a scene from the middle of *Metamorphoses*, where Ovid imagines a witchy version of his own creative alchemy in the form of Medea's cauldron of youth. The main guinea pig here is her father-in-law Aeson, who needs restoring, like Tithonus. But first the olive branch with which she stirs her regenerative brew serves as an accidental tester. On touching the liquid, it sprouts again, exchanging over-ripe fruit and dry bark for green leaves – autumn for spring:

> ecce uetus calido uersatus stipes aeno
> fit uiridis primo nec longo tempore frondes
> induit et subito grauidis oneratur oliuis;
> at quacumque cauo spumas eiecit aeno
> ignis et in terram guttae cecidere calentes,
> uernat humus floresque et mollia pabula surgunt.
> quae simul ac uidit, stricto Medea recludit
> ense senis iugulum ueteremque exire cruorem
> passa replet sucis; quos postquam combibit Aeson
> aut ore acceptos aut uulnere, barba comaeque
> canitie posita nigrum rapuere colorem,
> pulsa fugit macies, abeunt pallorque situsque,
> adiectoque cauae supplentur corpore rugae,
> membraque luxuriant. Aeson miratur et olim
> ante quater denos hunc se reminiscitur annos.

Met. 7.279–93

She stirred it all with the dry branch of a fruitful olive, mixing the depths with the surface. Lo, the ancient stick turned in the hot cauldron, first grew green again, then in a short time sprouted leaves, and was instantly heavily laden with olives. And whenever the flames made foam spatter from the hollow bronze, and warm drops fall on the earth, the soil bloomed, and flowers and soft grasses grew. As soon as she saw

> *this, Medea unsheathed a knife, and cut the old man's throat, and letting his old blood out, filled the dry veins with the juice. When Aeson had absorbed it, partly through his mouth, and partly via the wound, the white of his hair and beard quickly vanished and a dark colour took its place. In a moment, his leanness disappeared, and his pallor and dullness of mind. The deep hollows were filled with rounded flesh, and his limbs expanded. Aeson marvelled, recalling that this was his self of forty years ago.*

Old Aeson gets his dipping and is miraculously fleshed out, his wrinkly skin plumping up in an instant. A soupy broth of trees and people is stirred at the pulpy, fluid centre of a book (*liber* = book / bark) which is itself constituted and re-constituted by the poet's infinitely replenishable material (*materia* = timber).

To return to my original questions: do trees in *Metamorphoses* represent a parallel or a foil to human lifecycles and what can they tell us about the teleology of the poem? According to the overview given above, Ovid's trees appear to track the rise and fall of cities and dynasties, not those of individuals. Their *telos* is not self-combustion nor decay. Neither is it endangered status, as artist Max Peintner once imagined for a group of trees preserved as exhibits in a stadium of the future, huddled like vulnerable rare beasts in a menagerie – a picture not so different from the amphitheatres of ancient Rome, with their wooded backdrops and wild animal hunts (*uenationes*).[52] For Ovid, it is humans who are vulnerable and threatened while his trees express endless regenerative potential, defying the gnarled texture and hampering rootedness of each individual specimen. Aeson needs a witch to rejuvenate him; he has one reprieve only. In nature, by contrast, trees get to perform their backwards miracles, their seemingly impossible renewals, every single year. As long as we let them, they go on, perennially. In Philip Larkin's words:

> Last year is dead, they seem to say, Begin afresh, afresh, afresh.

Notes

1. Repeated by e.g. Sharrock 1994: 122 and Gowers 2005: 337. Text of *Metamorphoses* is taken from Tarrant 2004.
2. On the metaphor *bracchia*, arms/branches, see Perutelli 1985.
3. Hunt 2016: 198.
4. Pliny, *HN* 17.224 (arthritis), 17.248 (hair); cf. Col. *Ag.* 3.10.11 for an extended analogy between tree and human bodies.
5. Cf. P. Evans, 'Long live trees', *The Guardian* 29 December 1999: 'There is an inextricable link between people and trees, especially old trees.'
6. See Sharrock 1996: 119–22 for a sensitive discussion of the slippage of terms in Myrrha's transformation in *Met.* 10.
7. Gowers 2005.
8. Kohn 2013; Wohlleben 2016.
9. Hunt 2016: 173–223 discusses the debate around tree animation in antiquity; see Holmes 2016 on Greek philosophical debates.

10. Ov. *Met.* 3.441.
11. Jones and Cloke 2002: 4.
12. Coo 2007.
13. Pliny, *HN* 16.6.
14. Ov. *Met.* 3.80.
15. Ov. *Met.* 3.93–4. Cf. Butler in this volume on the inanimate and animate agents of the verb *gemere*.
16. Jones and Cloke 2002: 48.
17. Stein 2013: 94–107.
18. Payne 2016.
19. See Martelli in this volume on how Ovid's incorporation of multiple species with varying lifecycles and feeding habits disrupts neat chronological divisions; instead, 'knots of time' tie birds, for example, both to their evolutionary lineage and to the seasons of their prey.
20. Cole 2008: 97–9.
21. Cole 2008: 90.
22. Haeckel 1866: 2.300; cf. Gould 1977: 76–85.
23. Cole 2008: 80–100.
24. Cole 2008: 93.
25. See Sharrock 1996: 106–8 on synchronicity versus diachronicity in human–tree transformations.
26. Hunt 2016: 46–54, 173–223.
27. See also Jacques 2013: 179 n. 10: dryads can be 1) semi-divine beings associated with trees and woodlands; 2) nymphs whose lives are coterminous with trees; 3) nymphs who inhabit trees.
28. Hunt 2016: 190–6: e.g. schol. *ad* Apollon. Rhod. 2.479: hamadryads are born and die *hama tais drusi* ('along with their trees'); cf. Serv. *ad Ecl.* 10.62: 'hamadryads are nymphs who are born and die along with their trees.'
29. 'Deep-bosomed and mountain-couching' (257) might indeed mean that they have attractive cleavages, as Olson 2012: 261–2 alleges; but surely the point is that their bodies follow the contours of mountains and valleys, shaping the landscape as both feminized and sexualized, as a goddess of love would see the world.
30. See Bergren 1989: 8–29 on the role of clothing, perfume and other surface aids in veiling and hinting at the goddess's own hidden sexuality.
31. Sharrock 1996: 119–22.
32. Feldherr 2002: 172–4.
33. See below on the reappearances of the laurel and evergreen foliage in *Met.* 15.
34. A more obvious allusion is to the simile at Hom. *Il.* 6.146–9, comparing the generations of men to those of leaves.
35. Nugent 1989.
36. Gowers (2005: 337) calls Book 8 'Ovid's *Silvae*'.
37. I prefer *trabes* (which Tarrant concedes may be correct) to his *faces*.
38. See also Zatta 2016: 107–9 for a reading of this episode.

39. See also Sharrock 1996: 119–22.

40. Sharrock 1996: 119–22.

41. Varro *ap.* Gell. *NA* 16.16: *non ut hominis natura est, sed ut arboris.*

42. Gregorić 2005.

43. Chambers 1961.

44. Detienne 1972: 76–7.

45. *Met.* 12.514–21.

46. Littlefield 1965.

47. Backdating Jupiter's prophecy at Virg. *Aen.* 1.291: *aspera tum positis mitescent saecula bellis*, 'then, with wars at an end, harsh ages will grow mellow'.

48. Littlefield 1965: 466: 'Naturalization of Aeneas, metamorphosis of foreign prince into native-born divinity, roots the work and its purpose firmly in the rich Italian soil.' See also Gowers 2011.

49. Hunt 2010: 55.

50. A god's own metamorphosis into a tree is rare, presumably because it would limit movement and agency (as the anonymous referee suggests, the combination of steadiness and ephemerality represented by a tree gives it a 'privileged' relationship to human beings). I can think only of *Met.* 11.244 (Thetis eluding Peleus) 'then you became an earthbound tree: Peleus clung to the tree' (*nunc grauis arbor eras: haerebat in arbore Peleus*), which may recall Hor. *Sat.* 2.3.71–3 (Proteus, Thetis' teacher in escape artistry, as metaphor for an evasive debtor): 'he will become a boar, then a bird, then a rock, and if he wants, a tree' (*fiet aper, modo auis, modo saxum et, cum uolet, arbor*).

51. *Ann.* 13.58. He would have enjoyed the tale of the oak-tree President Macron gave to President Trump in 2018 to plant in the White House as a symbol of Franco-American friendship, but which has already died.

52. Peintner 1970/71.

SCIENCE/WISDOM TRADITIONS

CHAPTER 6
THE WORLD IN AN EGG: READING MEDIEVAL ECOLOGIES

Miranda Griffin

The Middle Ages inherit many Ovids from antiquity.[1]

The ecology that Ovid articulates in the *Metamorphoses* may have been particularly appealing to the medieval environmental imagination because its intertwining narrative strands invite – if not compel – the reader to participate in a vision of the world as a series of perspectives fluctuating between human and nonhuman, animal and divine. As it does so, the *Metamorphoses* asks insistent and troubling questions about where these categories begin and end, and why these categories matter. While appearing to conduct the reader from the beginning of time to the flourishing of Rome, the narrative of the *Metamorphoses* flits between speakers and time frames, prompted by patterns of similarity and memory. This essay explores these patterns of similarity and memory as they underpin structures of knowledge about what medieval European scholars understood as the created world, humanity's place in it, and its status as a book to be interpreted and learned from in order to lead the best possible Christian life.

To imagine the world in this way relies on the identification of patterns of resemblance from the infinitesimal details of God's work to the ineffable vastness of the universe. The figure of resemblance on which I focus in this essay is the egg – a figure whose its shape and layered composition made it available for comparison with the stratified elements of the creation of the universe. As I will show, through a creative (mis)reading of Ovid's name, reiterated across medieval traditions glossing the *Metamorphoses*, an association is forged in medieval manuscript culture between Ovid and the representation of the cosmos as an egg. This essay focuses on the representation of the cosmic egg in text and image of two distinct yet entwined traditions of medieval wisdom literature – or *science*, as it was called in medieval French: moralized readings of the *Metamorphoses*; and the bestiary tradition. Both traditions, I propose, represent ecology as a matter of interconnections and similitude, where the human and the nonhuman are situated in complex hermeneutic networks.

In order to analyse this world view more closely, I refer to the work of the French philosopher of science, Bruno Latour. Latour's work emanates from his thesis that a modern insistence on dividing life into rigid categories such as nature and culture results in an understanding of the world that is destructive and wilfully neglectful of the nonhuman environment. In *Facing Gaia*, a series of lectures on ethics and ecology, Latour calls on figures of metamorphosis in order to envisage the 'dizzying otherness of existents'.[2] Metamorphosis, then, is key to Latour's account of Gaia – his term, borrowed from Lovelock, to describe the ever-shifting networks of being and power which characterize our understanding of the cosmos and our place in it. For Latour, the metamorphosis of Gaia stands as a counter to the totalising notion of the Globe, a reductive vision of the world held apart from human action, as if Nature is a separate entity on which humans can gaze from a godlike perspective.[3] While Latour characterizes such global thinking as akin to medieval theology, I want in this essay to propose that the textual and visual representations of creation on the pages of manuscripts transmitting medieval *science* share a profound interest in the dizzying otherness of existents.

The world of the *Ovide moralisé*

The influence of Ovid's *Metamorphoses* on European perceptions of matter and its mutability was profound from the twelfth century onwards, as Caroline Walker Bynum has demonstrated.[4] While Bynum argues that Aristotelian conceptions of change were predominant in the thirteenth and fourteenth centuries, the *Metamorphoses* remained an important source for medieval French authors and artists. This is evident from the great success of the extraordinary, compendious, anonymous work, the *Ovide moralisé*, which dates from the first half of the fourteenth century.[5] This was the first full-length translation of the *Metamorphoses* into French; it also amplified Ovid's poem vastly by interpolating a series of moralisations of the stories, reading each of Ovid's tales as allegories about world history since the creation, Biblical exegesis, and the best way to lead a pious Christian life. The *Ovide moralisé* formed part of a tradition of medieval wisdom literature for which the created cosmos was a text to be interpreted; humanity's place in the environment was understood as facilitating an approach to an understanding of the divine, as well as a concomitant acknowledgment that no such understanding could ever be complete. bestiaries were also part of this medieval tradition: composed in many languages, including Latin, Old English, German and French, and hugely popular in the Middle Ages, bestiaries list animals – both familiar and fantastic – and give accounts of their behaviour (or '*nature*'), portraying them in both words and images, and then gloss that behaviour in ways which explain and guide human activity.[6] The accounts of animals' '*nature*' and the allegorical meanings allocated to them derived from the Greek *Physiologus* (second century CE), which gathers traditional material about animals from a wide range of sources, including those which inform the *Metamorphoses*: as we will see later in this essay, Pythagoras' sermon in *Metamorphoses* Book 15 refers to descriptions of animals which resonate with those in bestiaries and derive from the same sources.

The medieval tradition of *science* in which the *Ovide moralisé* and the bestiary tradition participate holds that that the world is, as the twelfth-century sage Hugh of St Victor proposed, a book written by God, the individual creatures within it being figures to be interpreted.[7] This tradition, then, anticipates by several centuries (which is, of course, but the blink of an eye in cosmological time), the ecocritical project of analysing the cosmos as a space in which 'meaning and matter are inextricably tangled, constituting life's narratives and life itself.'[8] The trope of the Book of Nature is pervasive in medieval literary culture,[9] and includes learning inherited by medieval European Christian *science*, as the *Ovide moralisé* makes clear from its opening lines:

Se l'escripture ne ment,
Tout est pour nostre enseignement
Quanqu'il a es livres escript,
Soient bon ou mal li escript.

OM 1.1–4

If scripture does not lie, everything is for our education when it is written in books – be that writing good or bad.

All written matter, then, should be considered as appropriate material for the sort of interpretation which the *Ovide moralisé* so comprehensively models – a meticulous allegoresis which subjects its source text to readings that lead to a pious understanding of the composition of the cosmos; the tenets and ethics of Christian life; and the biblical narratives that articulate them. And this 'Tout' includes – daunting though they may be to this programme of reading – the scandalous, amoral, non-Christian tales of the *Metamorphoses*, and the environment teeming with mutating forms and sexual violence in which they unfold. The living environment and the books which transmitted wisdom about it are, then, figures for one another in the Middle Ages. Writing, in this worldview, is part of the environment, and the environment is an inscription by the hand of God.

In the following section, I investigate the figure of the cosmic egg and its identification with Ovid in the Middle Ages through a detailed examination of the opening folios of two manuscripts which transmit the *Ovide moralisé*. I then discuss another manuscript which contains two comparisons of the world to an egg, one of them in the context of an extensive bestiary. With reference to Latour's evocation of figures of metamorphosis, I argue that these manuscripts propose an environmental imagination which trace vibrant networks of resemblance, enabling us to answer Latour's call to 'retrace other cosmologies,'[10] and to think about the world in the 'metamorphic zone'.[11] The metamorphic zone, in geology, is evident in rock formations which have been subjected to intense forces: it is therefore a visual trace of extreme physical processes detected in solid, static, stone. Latour invites us to imagine ourselves in that zone – as does Ovid, via his vivid descriptions of transition between human and nonhuman embodiment. In the concluding section, I return to the *Metamorphoses* and to the medieval treatment of its depiction of hearkening to a learned disquisition on the twists, turns, and surprises of the environmental imagination: Pythagoras' sermon.

Manuscripts and the metamorphic zone

The cosmic egg – the notion that the cosmos is a made up of a series of layers which can be compared to the layers of yolk, albumen, membrane and shell of a humble egg – is a figure which dates from at least a century before Ovid composed the *Metamorphoses*.[12] It was attributed in the pseudo-Clementine *Recognitions* to Orpheus, rather than Ovid, in an account which depicts Orpheus singing of the division of chaos into elements at the beginning of the Earth.[13] The association between Ovid and the figure of the cosmic egg is traced by Frank Coulson to the *Accessus*, the introduction to the thirteenth-century Latin 'Vulgate' commentary on the *Metamorphoses*, an important source-text for the *Ovide moralisé*. In the *Accessus*, the name 'Ovidius' is glossed as deriving from 'ovum dividens'.

> Ouidius autem
> nomen est proprium et ethimologizari potest sic: Ouidius
> enim dicitur quasi 'ouum diuidens'.[14]

The egg and the world thereby become metaphors for one another; through this medieval wordplay, the connection between them is assigned to the authority of Ovid.[15] Underpinning this understanding of the universe is the medieval apprehension of the ancient structure of thought which insists on images of similitude and equivalence between the unthinkably enormous (the star systems of the firmament), the unfathomably divine (God's plan of the universe), the insignificantly mundane (an egg) and the vanishingly tiny (elements which make up the world). The commonest form of this image is the microcosm, where the human individual is understood as a miniaturised yet complete model of the cosmos. As Kellie Robertson observes in an article about the didactic and epistemological mechanisms of medieval microcosmic thinking, 'the space between body and world is at once an impossible distance and an invitation':[16] the intuition that the human body and the cosmos function analogously in spite of their different scales leads to the concept that they are part of a divine plan which operates on the basis of comparable patterns across space, time, and scale. The body in this schema is understood as a *minor mundus*, a little world.

The figure of the cosmic egg also casts the egg as a *minor mundus*. Crucially, however, the image of the world as an egg offers not a reading of the cosmos onto the human body, but a microcosm of the nonhuman. In this sense, it decentres the human from the patterns of resemblance which structure creation, although it calls on the perspicacity of human readers to discern this resemblance.

The moralizations of the *Ovide moralisé* depend upon these patterns of resemblance. Its first book draws repeated parallels between the representations of the creation of the world in the first book of the *Metamorphoses* and the first book of the Old Testament, Genesis. In the *Metamorphoses*, Ovid describes the process, overseen by an unnamed pagan deity, of separation of the elements, and their arrangement in concentric layers according to their weight:

> ignea convexi vis et sine pondere caeli
> emicuit summaque locum sibi fecit in arce;
> proximus est aer illi levitate locoque;
> densior his tellus elementaque grandia traxit
> et pressa est gravitate sua; circumfluus umor
> ultima possedit solidumque coercuit orbem.
>
> *Met.* 1.21–31

The weightless fire, that forms the heavens, darted upwards to make its home in the furthest heights. Next came air in lightness and place. Earth, heavier than either of these, drew down the largest elements, and was compressed by its own weight. The surrounding water took up the last space and enclosed the solid world.

The *Ovide moralisé* author interprets Ovid's description as analogous to the division of light and dark, the creation of the firmament, and the separation of land and sea detailed in Genesis: indeed, there is little distinction between the stories of creation presented in the section of the first book of the *Ovide moralisé*, which translates Ovid's account (*OM* 1.147–98), and the later section, which moralizes it (*OM* 1.341–86). Between these two accounts, we find, unexpectedly inserted into the otherwise reasonably faithful translation of Ovid's verse,[17] an attribution of the cosmic egg figure to Ovid.

Pour manifester clerement,
Et pour donner entendement
Coment vait li ordenemens
Et l'assise des elemens,
A ce veoir nous avisa
Ovides, qui l'oeuf devisa,
Si vault similitude faire
Tel, qui le nous monstre et desclaire
Apertement, si com je cuit:
C'est par un oeuf en quoque cuit.

OM 1.199–208

In order to demonstrate clearly and to convey understanding of the process of ordering and positioning of the elements, and in order to envisage this, Ovid, who described the egg, wished to make a figure of similarity thus: I believe he shows and states clearly that it is understood by a boiled egg in its shell.

It is worth pausing to note here the striking use of rhyme in this description: never one to resist an equivocal rhyme, no matter how bizarre, the *Ovide moralisé* author manages here to set up a rhyme in 'cuit' ('je cuit', I believe; 'un oeuf en quoque cuit', a boiled egg). In the examples of the cosmic egg figure that I discuss in this article, no other goes to the effort of specifying whether or not the egg is cooked. While the egg is evidently a figure for birth and regeneration – themes found in abundance in the opening book of the *Metamorphoses* and the first book of the *Ovide moralisé* – here it is presented as an egg to be eaten.

The *Ovide moralisé* author then elaborates that the yolk, the white, the membrane and the shell of the egg are interpreted respectively as earth, water, air, and fire, in a scheme corresponding to the series of concentric circles of elements described in the *Metamorphoses*.

En l'oeuf, ce me samble, a trois choses
Qui sont dedens la quoque encloses:
Le moieuf, l'aubun, la pelete,
Qui plus est près de la quoquete.
Le moieus nous note la terre,
Qu'einsi com li aubuns l'enserre,
Par cui nous devons la mer prendre,
Tout ensement doit l'en entendre
Que la terre est avironnee
De mer. Aprez est ordenee
La pelete tenve et deugie,
Qui sor ces deus est assegie :
Tout ensement vault Dieus former
L'air moiste sor terre et sor mer.
Aprez vient par ordenement
La quoque, qui l'estendement
Dou ciel nous represente et note.

OM 1.209–225

In the egg, it seems to me, there are three things enclosed in the shell: the yolk, the albumen and the membrane, which is closest to the shell. The yolk represents to us the Earth: it is enclosed in the albumen, which we should understand as the sea. This should all be understood together, since the earth is surrounded by the sea. After this the fine, delicate membrane is arranged, placed on top of the other two: in this way, God wished to form moist air over land and over sea. Next in this arrangement comes the shell, which represents and shows to us the establishment of the heavens.

The image of the egg, then, evokes the layered elements as they are described by Ovid: more than simply a metaphor adduced by Ovid, the cosmic egg becomes closely associated with this author, via the playful reading of his name.

Illustrations in three surviving manuscripts of the *Ovide moralisé* emphasize this association by representing Ovid holding an egg: Rouen, Bibliothèque municipale 0.4 (fourteenth century); Copenhagen, Kongelige Bibliotek, Thott 399; and Paris, Bibliothèque Nationale de France, fonds français 137 (both fifteenth century). Where Thott 399 transmits the verse version of the text, BnF f fr 137 has a prose version.[18] I focus on the two fifteenth-century manuscripts in this essay, since their mise-en-page is richly indicative of the environmental imagination of the scribes and illustrators who produced them.[19] Thott 399 was produced in Northern France between 1454 and 1460 and illustrated by an artist known as the Master of Rambures. BnF f fr 137 is the only manuscript of the prose version of the *Ovide moralisé* to contain illustrations: they are the work of an artist known as the Master of Margaret of York; and Françoise Clier-Colombani discerns correspondences between his illustrations and those of the Master of Rambures. BnF f fr 137 was produced in Bruges in the 1470s, was owned by the fifteenth-century bibliophile Louis de Bruges, and became part of the royal library of King Louis XII of France.[20]

Figure 1 Paris, Bibliotheque Nationale de France fonds français 137 fol 1.

The opening folios of each of these manuscripts' versions of the *Ovide moralisé* affords what could well be described as a vision of what Latour calls the 'dizzying otherness of existents'.[21] BnF f fr 137 (Figure 1) depicts God on the right of the image, in the midpoint of a cross formed by four frames, each one of which encloses an image of creation: clockwise from top left, we see stars and birds in the firmament; fire enclosing a creature (probably a salamander, which was thought to live in fire); fish of many forms swimming in the sea; and a human figure, Adam, kneeling and naked surrounded by trees and rocks, with some animals in the background which he may be in the process of naming. The images on the left and right of the folio are separated by a white frame, which is perhaps supposed to indicate an external wall of a building: half-way up this division there is what looks like an architectural feature of a figure under an intricately carved canopy. On the left-hand side of the folio, enclosed by this wall, beneath what looks like a very elaborate balcony, and against a richly decorated backdrop whose colours recall the starry firmament on the right-hand side of the page, a man sits on a wooden dais. He turns away from the book propped up on a lectern towards a group of men. In his left hand he holds an egg delicately between his fingers; he points to it in an expository, didactic fashion with his right hand as the men in the audience lean forward to listen to him. This learned man, we are to understand, is Ovid, calling on the figure of an egg, which echoes his name, to explain the composition of the earth. Beneath this composite image, the first few lines of the *Ovide moralisé en prose* have been copied; around the images and the text is another frame, in which accurately depicted plants and animals of identifiable species are mingled (we can easily identify strawberries, grapes, a porcupine, and a peacock, as well as the more stylised acanthus leaves).

Folio 26r of manuscript Thott 399 (Figure 2) also depicts Ovid as a pink-robed sage holding an egg. In this image, Ovid sits within a building whose wall has been removed for us to see inside, alone at a table, reading and holding an egg in his right hand. Outside, fallen angels rain down around him, metamorphosing into demonic forms and plunging towards the fires of hell. This scene illustrates the beginning of the moralisation of Ovid's description of the separation of the elements, in which the *Ovide moralisé* author describes the fall of the proud angels immediately after God's creation of heaven and earth (*OM* 1.364–8). In the image in Thott 399, God looks down on the tumbling angels and Ovid's contemplation from a heavenly vantage point, against a background of gold leaf. A red semi-circular frame and the clouds that surround it suggest that God is positioned outside of the material cosmos he has created: this cosmos is echoed in the transparent sphere that he holds in his left hand, reflecting the opaque egg held by Ovid in his shelter. The opening lines of the *Ovide moralisé* are themselves framed by glossing, some of which spills over the marginalia depicting hybrid creatures standing on small grassy islands against a field of abstract gold swirls on a dark background.

Martha Dana Rust observes that the technicalities of writing are always presented by medieval scholars as rich in metaphorical possibilities: if the created natural world can be understood as a book, she argues, then a book can be understood as constructing spaces in which the created world can be experienced, scrutinized and interpreted.[22] The folios of the Paris and Copenhagen manuscripts, then, can be read as ecologies, in which

Figure 2 Copenhagen, Kongelige Bibliotek, Thott 399, fol 26r.

the human and the nonhuman are presented in a set of intricate interrelations across space and time. In the Paris manuscript, the elements that make up the cosmos are depicted framing and containing human and nonhuman figures. These human figures are shown in moments of apprehension of the nonhuman: Adam names the animals, Ovid explains the world with reference to the egg. Indeed, both Adam and Ovid might be seen here as reading the book of nature. This ecology also integrates divergent timeframes: on the right, the creation story articulated in Genesis is depicted as a series of originating moments; on the left, these moments and the world whose inception they chronicle are enclosed in an egg and subject to the learned exposition of the sage.[23] But we can also read the process of interpretation from left to right, as Ovid first describes the make-up of the earth of the left, and then the *Ovide moralisé* expounds this description in terms of the Genesis creation story on the right. In the Copenhagen manuscript, the exchange between pre-Christian past, moralising present, and divine eternity is differently configured, as the chaotic bodies of the metamorphosing angels not only represent the results of sin in Genesis, but also the pre- Christian narratives the *Ovide moralisé* takes as its source-text. Ovid may, in his pink robes and hat, be dressed as a late medieval man of letters, just as his audience in the Paris manuscript is clothed in recognizably medieval outfits, but he is also being represented in this image as the Classical *auctor*, the writer of a pre-Christian text which can nevertheless be interpreted as a guide to Christian wisdom and life.

The more serene body language of God and Ovid conveys a perspective from which to contemplate this disruption – Ovid is situated in a rocky landscape, while God surveys the created world and holds a shimmering globe to echo both that creation and the humble egg held by Ovid. Both God and Ovid might seem, in this image, to be subscribing to the view of the Globe against which Latour warns in *Facing Gaia* – after all, as he says, 'he who looks at the Earth as a Globe always sees himself as a God'.[24] But this extraordinary illustration is clearly a fantasy, taking place in the space of a manuscript page but conjuring a location to which no mortal can have access. It proposes an imagined environment teeming with patterns of resemblance and interpretation in the human and nonhuman world, inviting a reader to trace resonances between and across text and image.

A profound ontology

A popular medieval cosmogram, the cosmic egg figure calls on notions of scale and epistemology based on perceived structural resemblance – rather than, say, a taxonomy of species or genetics which might govern modern classification and analysis. This medieval worldview chimes with Latour's challenges to modern ways of understanding the world, challenges which can often call on figures of metamorphosis. Latour capitalizes on the status of the modern and modernity as a mode of thought rather than a simple chronological phenomenon. As is evident from the title of his 1991 work, *We Have Never Been Modern*, Latour sees the Modern insistence on sectioning off intellectual disciplines which guide enquiry into the lived environment as harmful. The division by these disciplines of the world into the gratuitous categories Nature and Culture has, he argues, 'de-animated' the world.[25]

> The idea of a Nature/Culture distinction, like that of human/nonhuman, is nothing like a great philosophical concept, a profound ontology; it is a *secondary stylistic effect*, posterior, derived, through which we purport to *simplify* the distribution of actors by proceeding to designate some as animate and others as inanimate.[26]

Instead of this claustrophobic, stultifying binary, Latour proposes that the world be understood as a 'metamorphic zone', a geological term designating rock formations which indicate processes of pressure or heat: this is a metaphor, then, which refers to something as permanent as rock yet depicts it as lively and mobile, deployed to trace leaps of agency between bodies and forms. It is a space of thinking which, in the face of dramatic global change, 'is leading us, little by little, beneath and beyond the superficial characterizations, to a radically new distribution of the forms granted to humans, societies, nonhumans, and divinities'.[27] A zone, then, in which one might sing of bodies transforming into new forms, or vice versa.[28]

Metamorphosis is a favoured trope for Latour. In *An Enquiry into Modes of Existence*, Latour evokes the 'Beings of Metamorphosis.'[29] The power of these beings, Latour argues, lies in the metamorphic force of imagination: they enable us to envisage our pasts and presents, to imagine ourselves otherwise; 'they can transform us at any time'.[30] And in his meditation on the Covid-19 pandemic, *Où suis-je?*, Latour takes as his starting point Kafka's story, *Metamorphosis*, whose hero has been transformed, and awakes to a strange confinement, needing to reassess his bodily dimensions, and constantly wondering where he is.[31] As a hermeneutic figure of thought, metamorphosis offers a means of seeing more than one form at once, of apprehending the world as a space of flux, dynamism, and potential, where 'actors, with their multiple forms and capacities, never stop exchanging their properties'.[32] There are resonances between this vision of the world and the environmental imagination proposed by medieval scalar figures of thought such as the cosmic egg, which, Robertson reminds us, hold disparate bodies simultaneously present: the scalar metaphor, she points out, 'is a move from the realm of resemblance to the realm of being, from likeness to identity, from imitation to participation'.[33] Similarly, Latour's recurrent figure of metamorphosis demonstrates a capacity to think across forms, to see bodies developing into and out of one another, from one movement and one moment to the next, to observe surprising resonances and resemblances between apparently dissimilar entities, and to revel in these resemblances as figures for thought which enable us patiently to plot the networks of our environment. For Latour, we can all be Proteus and Morpheus, fluctuating in our imaginations through evolving forms: 'This proteiform character is familiar to all of us,' Latour observes, emphasizing the central role of the Beings of Metamorphosis to our psychic realities. 'What would we do without them? We would be always and forever the same.'[34] It is in this metamorphic zone that we might see medieval receptions of the *Metamorphoses* functioning, in their emphasis on learning from bodily transformation, their focus on scalar forms where an egg can stand for the world, and the complexity of the manuscript pages which transmit them.

The world, the egg and the soul

Dated 1267/8, manuscript Paris, Bibliothèque de l'Arsenal 3516 was produced around fifty years earlier than the composition of the *Ovide moralisé*. Made in Northern France, it is a compendious manuscript, gathering together a wide range of texts: Claudia Guggenbühl has shown that these are organized into categories pertaining to biblical narratives, vernacular literature, scientific texts, and treatises about arms and armour.[35] This manuscript, then, gives a composite, and sometimes contradictory, vision of the world. In the terms of the *Ovide moralisé*, the texts it transmits give a compelling account of the 'Tout', the entirety of the created environment, whose literary depiction is there for our education. Amongst these are two works which describe the cosmic egg: an extended version of the bestiary, originally attributed to Pierre de Beauvais;[36] and a *Mappemonde*, a geographical description of the world, in a version unique to this manuscript.

The two written instances of the cosmic egg in Arsenal 3516 occur on folios 157r (column 1) and 211r (columns 1–2) and use identical wording.

> *Li mondes est une cose tot ensamble qi est de ciel et de terre et des .iiii. elemens*
>
> [...]
>
> *La forme du monde puet on veoir en .i. oef: li moiels est la terre; li aubuns est li blans, qui avirone le moiel, et est com li airs entor la terre; l'escaille qui est entor par defors est li firmamens qui tot enclot, et air et terre.*
>
> *The world is a thing which adheres together: it made of sky, earth and the four elements.*
>
> [...]
>
> *The form of the world can be seen in an egg: the yolk is the earth; the albumen is the white which surrounds the yolk, and is like the air around the earth; the shell which is around the outside is the firmament which encloses everything, both air and earth.*

In the *Mappemonde*, the cosmic egg is described as a preamble to the division of the world not into elemental layers but into continents.[37] The version of the bestiary transmitted by Arsenal 3516 embeds the figure of the cosmic egg in a section entitled 'De coi li home est fais et de sa nature' (On the composition and nature of man). The human body is presented in this context, then, not as a microcosm of the created world, but as just another creature within it, yet another chapter in the bestiary, to be described and moralized alongside other beasts. Indeed, the image accompanying this section (Figure 3) situates a naked human figure standing next to a tree framed by the four elements and the beasts that live in them – the salamander (depicted in this image as a sort of winged hybrid creature, looking a little like a dragon or a griffin) in fire; the chameleon (a creature understood in bestiary tradition as a bird) in air; the herring in water; and the mole in earth. Humanity is another mode within this ecology, a point at which the elements

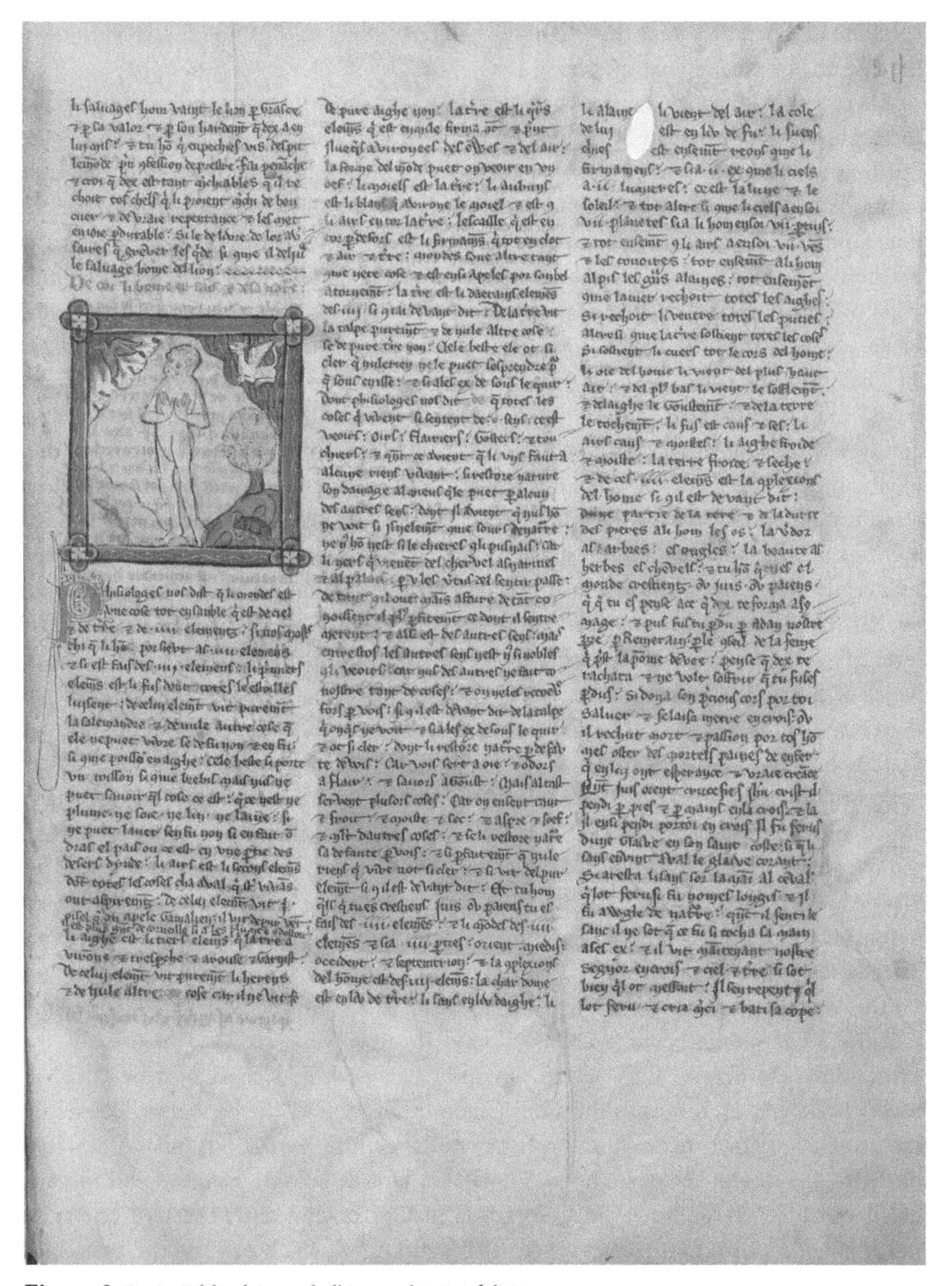

Figure 3 Paris, Bibliothèque de l'Arsenal 3516, fol. 211r

combine in the same way as they do in the composition of the cosmos as it is outlined in the opening of the *Metamorphoses*.

There are striking visual echoes between the image accompanying the account of man in the bestiary of Arsenal 3516 and the right-hand panel of the illustration on the opening folio of BnF f fr 137. Although these manuscripts come from different ages, locations and traditions, their images of the elements and the creatures that dwell in them indicate a similar environmental imagination, in which the created cosmos is harmoniously quadripartite; and both images are complemented by text comparing this composition, depicted visually as four juxtaposed equal shapes, to the concentric layers of an egg. Whereas the figure that stands at the intersection of the four elements in Arsenal 3516 is a naked man, in BnF f fr 137, it is the flying, haloed figure of God; the naked man is shifted to the bottom right-hand quadrant, becoming Adam, kneeling amidst the trees of Eden. This vision of an ecology which integrates humanity within a vibrant enmeshment of actors resonates with Latour's insistence that modern classification of the lived environment is contingent: it recalls his insight, which I quoted above, that the Nature/Culture split is rhetorical rather than ontological, and that modernity's trick has been to efface that distinction. Latour's attack on modernity and the harm it does to the ways we can understand our environmental existence is precisely levelled at this rhetorical move rather than primarily addressing the historical period in which it plays out (although the latter is the effect of the former), and he does not therefore identify a historical period whose articulation of knowledge about the world is preferable to modernity. Scholars of pre-modern culture have pointed out that Latour's approach can be productive when exploring notions (like the cosmic egg or the bestiary) which appear hopelessly muddled to modern science. This is apparent from the title of the introduction to a collection of essays on Latour and medieval literary culture, 'We Have Always Been Medieval'.[38] In this collection, Emma Campbell observes that 'medieval bestiaries exemplify a premodern approach to ontology that Latour's Moderns have cast aside in their understanding of the natural world'.[39]

The version of the bestiary in Arsenal 3516 is an unusually extended one, and includes some animals which are not found in other versions. One of these is the extraordinary orphenay bird, which has the head, neck and chest of a peacock, an eagle's beak, and lays its eggs on water. The moralization of the orphenay focuses on the prophetic knowledge mother orphenay birds seem to have about the moral worth of their chicks before they are born. 'Li oisel set tant de sens de nature de lui meisme qui il voit bien et conoist le quel oef ara le meillor pochin ens' (The bird knows so much about its own nature that it sees well and knows which egg has the best chick inside it).[40] When the orphenay lays her eggs, those containing good chicks, that the mother will love, stay near the top of the water; but those containing inferior, unlovable chicks, sink to the sandy bottom of the sea. The good chicks hatch and stay under their mother's wing, and she guides them joyfully to their father. But those who hatch beneath the waves spend all their lives in the shadows. Unlike some of the other interpretations of bestiary animals, the moralization of the orphenay does not come as much of a surprise:

> Tu, hom crestiens, chi dois tu oïr et entendre. La mer nos senefie cest mortel siecle. Li oisel est example de nostre mere sainte Glise. Li oef qui est en la mer nos senefie l'ome qui est en cest monde. Li oes est examples de l'home et li pochins qui est en l'uef est example de l'ame.[41]
>
> *You, Christian man, should listen and attend to this. The sea signifies this mortal world. The bird stands for our mother the holy Church. The egg in the sea signifies to us man in this world. The egg stands for man and the chick in the egg stands for the soul.*

In interpreting her own eggs, the orphenay is thus not only an extraordinary creature, but herself a reader of the book of the world, using her own intuition of the '*nature*' of her species and the individual iterations of it in her chicks. In the section of the bestiary on the nature of man, the egg and the human body are used as microcosmic images to aid the comprehension of the composition of the world. But in the entry for the orphenay, the egg now stands for the human body encasing the soul; the beginning of earthly life stands for the beginning of eternal life. As the good souls are raised to heaven, then the bad ones sink, rather like the falling angels who tumble around Ovid in the Thott manuscript. In the bestiary the shape of an egg represents the space of the world, but also the gravity and temporality of human life. This is an environmental imagination which articulates its own imaginative reading, proposing itself as a book to be read or an egg to be hatched. In the metamorphic zone of the manuscript page, the orphenay is imagined as an active reader of her own existence, and proposes a model of reflection to the readers of the bestiary, encouraging them to understand their own morality and embodiment as participating in a network of agency and interpretation, figures among others populating the Book of the World.

Eggs in the environment of the *Metamorphoses*

While Ovid does not make a comparison between the layers of an egg and the world, whatever the first book of the *Ovide moralisé* claims, in the final book of the *Metamorphoses*, the sage Pythagoras does weave together an exposition of the elements with a meditation on the extraordinary nature of eggs:

> Iunonis volucrem, quae cauda sidera portat,
> armigerumque Iovis Cythereiadasque columbas
> et genus omne avium mediis e partibus ovi,
> ni sciret fieri, fieri quis posse putaret?
>
> Met. *15.385–9*
>
> *Who would believe, if he did not know, that Juno's bird, the peacock, that bears eyes, like stars, on its tail; and Jupiter's eagle, carrying his lightning-bolt; and Cytherea's doves; all the bird species; are born from the inside of an egg?*

Figure 4 Copenhagen, Kongelige Bibliotek, manuscript Thott 399, fol 424r.

Just as Ovid is imagined by the *Ovide moralisé* as proposing the humble (boiled) egg as a strikingly banal figure for the whole cosmos, in Pythagoras's example, eggs stand for the incredible variety of the material environment, as astonishing as the tales of the *Metamorphoses* related throughout the rest of the poem. It is interesting to note that Pythagoras cites the peacock and the eagle here – both birds that combine in the holy hybrid figure of the orphenay.

Nestling in the richly decorative border of the first folio of BnF f fr 137, the peacock is an example of the variety of creation played out in the right-hand central image: the ecosystem of the manuscript here reinforces the ecology it represents. Ovid's description of the peacock likens it to the firmament, and the inclusion of the peacock in this opening folio might be read as an anticipation by the creators of this manuscript of Pythagoras' wisdom in the final book. In Book 15 of the *Ovide moralisé*, the birds mentioned by Pythagoras are interpreted as virtuous people, who rise above the sin of the world, rather like the good orphenay chicks rising to the surface of the sea. We see further cross-pollination of the traditions of *science* in Pythagoras' sermon: Pythagoras not only adduces the peacock as an extraordinary bird which seems as incredible as some of the stories in the *Metamorphoses*; he also refers to the traditions from which the medieval bestiary derived its material: bear cubs are born dead and licked into shape by their mother (*Met.* 15.379–81); the phoenix dies and is reborn in fire (*Met.* 15.391–406); and a hyena is alternately male and female (*Met.* 15.407–10).

Listening to Pythagoras' sermon is Numa, who at the beginning of Book 15 of the *Metamorphoses* has been chosen as the King of Rome: the opening of this book in Thott 399 is illustrated with his coronation, with the king and his subjects all arrayed as medieval nobles (Figure 4). Having listened in Crotona to Pythagoras, Numa returns, full of wisdom.

The *Ovide moralisé* renders this section as follows:

Numa moult ententivement,
Come sages et bien apris,
Ot tel enseignement apris
Et diligaument retenus,
Puis est en Rome revenus,
Plains de science et de savoir.

OM 15.1234–39

Numa learned this lesson attentively, like a wise, well-educated man; then he returned to Rome, full of learning and wisdom.

Echoing the 'enseignement' of the second line of the work, the translation of the *Ovide moralisé* presents Numa's attention to Pythagoras's sermon as a *mise-en-abyme*, a self-reflexive moment in the text in which a wise ruler hearkens to the teaching of a sage recounting tales of the world's diversity and flux – and appreciates that 'Tout est pour nostre enseignement.' Numa goes back to Rome full of 'science', just as we might imagine the men in the illumination on the first page of BnF f fr 137 will, having listened to Ovid's teachings on the world in the egg. As I mentioned earlier, BnF f fr 137 belonged to a King – as did several surviving manuscripts of the *Ovide moralisé*.[42] For these royal owners, the episode of a king learning the *science* about the wonders of the world would have been particularly powerful. The medieval garb worn by Ovid in the opening images of these manuscripts and by King Numa in the last book underline this equivalence – an analogy drawn not by manipulating variation in size, but by pointing out correspondences between present and past.

Ovid bequeaths to medieval literary tradition a vision of an interconnected world of human and animal, story and knowledge. In the *Ovide moralisé*, Ovid becomes a medieval sage, acquiring his own apocryphal pronouncement, based on a play on his name, about the structure of the world based on concentric circles which can be discerned in the heavens and in the layers of an egg. Just as the orphenay uses the egg to understand its own '*nature*', so the Middle Ages use Ovid to understand the cosmos: this hermeneutic gesture resonates with Latour's work, since it privileges the created environment as the object of study, detecting in it agency and animacy that are separate from the human activity so often placed at its centre. The process of reading enjoined by medieval receptions of the *Metamorphoses* is demonstrated by Latour's Beings of Metamorphosis: these are stories which give us access to the cosmos, as long as we have the patience to read.[43] And we can find the intellectual tools to intuit this through science, through *science*, and through the stories we tell ourselves about, through and within our environment. This integrated approach to understanding the relations between the human and the nonhuman on an ecological scale is something which we would do well to bear in mind in the third decade of the twenty-first century.

The closing words of *Où suis-je?* summarise this approach, resonating in a remarkable way with the extraordinary image that accompanies the opening of the *Ovide moralisé* in the Thott manuscript. Rather than thinking we can hold the globe, or an egg, in our hands, as Ovid and God are imagined doing, Latour enjoins us to embrace the metamorphic possibilities of living with the pulsating agencies that characterise our planet:

> Sous la voûte du ciel, redevenue pesante, d'autres humains mêlés à d'autres matières forment d'autres peuples avec d'autres vivants. Ils s'émancipent enfin. Ils se déconfinent. Ils se métamorphosent.[44]
>
> [Beneath the vault of the heavens, become weighty once more, other humans mixed with other matters form other people with other beings. Finally, they are emancipated. They emerge from confinement. They metamorphose.]

Notes

1. On the reception of the *Metamorphoses* in the Middle Ages, see Coulson 2011; Coulson 2007; and Desmond 1989.

2. Latour 2017: 36.
3. Latour 2017: 130–6.
4. Bynum 2011: 234–6.
5. For the *Ovide moralisé*, I refer to the edition by de Boer et al. (1915–38), hereafter referred to in parentheses as *OM*. On the influence of the *Ovide moralisé*, see Armstrong and Kay 2011: 80–90.
6. For the history of the Bestiary genre, see McCulloch 1962: 15–77 and Kay 2017: 7–11.
7. For discussion of Hugh's formulation, see Franklin-Brown 2012: 42–8 and Kay 2017: 28. On the relation between the *Ovide moralisé* and medieval encyclopaedism, see Ribémont 2002.
8. Iovino and Oppermann 2014: 5. See also Oppermann and Iovino 2017.
9. See Curtius 1953: 310–2.
10. Latour 2017: 37.
11. Latour 2017: 58.
12. See Dronke 1974: 79–99.
13. Dronke 1974: 84.
14. Coulson 1991: 25.
15. Mitchell 2014: 45–6.
16. Robertson 2019: 610.
17. On the translation technique of the *Ovide moralisé*, see Possamaï-Pérez 2006: 27–178.
18. On the prose text and its relation to the illustrative programme of BnF f fr 137, see Harf-Lancner and Pérez- Simon 2015.
19. For a description of the Rouen manuscript's representation of Ovid holding an egg, see Clier-Colombani 2015: 25–6.
20. Clier-Colombani 2017: 37–8, 53–6.
21. Latour 2017: 36.
22. Rust 2007: 14.
23. Clier-Colombani 2017: 56 describes the first folio of BnF f fr 137 in this way: 'Cette disposition traduit avec l'originalité d'un art consommé la volonté du peintre de faire découvrir au lecteur la mythologie antique à la lumière de la foi chrétienne, en disposant la figure d'Ovide comme vecteur – ou intermédiaire – entre la cosmogenèse païenne (partie gauche de l'image), et chrétienne (partie droite), et de montrer ainsi la puissance créatrice de l'esprit divin et la miséricordieuse attention qu'il accorde à sa création' (This arrangement translates with the originality of a consummate artistry the desire of the painter to present to the reader the ancient mythology in the light of Christian faith, using the figure of Ovid as a vector – or intermediary – between the pagan creation of the cosmos (on the left of the image) and the Christian one (on the right) and thus to show the creative power of the divine spirit and the merciful attention that he gives to his creation).
24. Latour 2017: 136.
25. Indeed, the second of his 2015 lectures collected in *Facing Gaia* is entitled 'How not to de-animate nature' (Latour 2017: 41–74).
26. Latour 2017: 68, original emphasis.
27. Latour 2017: 119.

28. For a discussion of the *Ovide moralisé*'s translation of the opening lines of the *Metamorphoses*, see Griffin 2012.
29. Latour 2013: 181–205.
30. Latour 2013: 205.
31. Latour 2021.
32. Latour 2017: 57.
33. Robertson 2019: 624–5.
34. Latour 2013: 201.
35. Guggenbühl 1998.
36. For detailed explanations of why Pierre de Beauvais cannot be the author of this long version, see Baker 2003 and the introduction to Baker 2010: 9–137.
37. It is used in a similar way in another *Mappemonde* text, this one authored by Pierre de Beauvais. Cf. Angremy 1983: 316–50 and 457–98, ll. 29–82.
38. Desmond and Guynn 2020: 1–7.
39. Campbell 2020: 130.
40. Arsenal 3516, fol. 212v; *Le Bestiaire*, ed. Baker 2010: 242.
41. Arsenal 3516, fol. 212v; *Le Bestiaire*, ed. Baker 2010, 243.
42. Griffin 2016.
43. See Latour 2013: 205.
44. Latour 2021: 165.

CHAPTER 7
THE TITANIA TRANSLATION: *A MIDSUMMER NIGHT'S DREAM* AND THE TWO *METAMORPHOSES*

Julia Reinhard Lupton

A Midsummer Night's Dream, written in 1595–96, is arguably Shakespeare's most Ovidian play: the story of Pyramus and Thisbe comes straight from *Metamorphoses*, Book 4; the love chases invert Apollo and Daphne, and the aetiology of love-in-idleness burgeons out of similar floral fantasies in Ovid.[1] The navel of the *Dream* is the metamorphosis of Bottom, at once an endlessly giving generator of asinine comedy and a bracing plunge into interpenetrating forms of life. The only realized animal transformation in Shakespeare, Bottom's translation interweaves aspects of Actaeon, Midas, the Minotaur and Io, while also borrowing from another *Metamorphoses*, *The Golden Ass* of Lucius Apuleius, a picaresque novel written in Latin by a Greek-speaking North African in the second century CE and translated into English by William Adlington in 1566.[2] Bottom's metamorphosis meets Titania's *amour fou*, an engineered infatuation with analogues in the obscene coupling of Corinthian women with the ass Lucius in Book X of *The Golden Ass*, as established by Jan Kott in his stunning reading of the play in *The Bottom Translation*.[3] Yet Shakespeare's fairy queen is also a divinity, identified with Diana as well as the demiurges of British folklore and sharing attributes with the Isis who restores Lucius to his human form at the end of Apuleius' novel. In *Midsummer*, Shakespeare interweaves the two *Metamorphoses* in order to stage a serio-comic mystery play in which manifold vitalities, cults, and cultures mix and mingle at the hybrid heart of his Orphic forest.

The allegorical tradition of the Middle Ages and the Platonizing mythopoesis of Renaissance humanism placed Ovid and Apuleius in an ancient wisdom tradition where ideas about nature from Jewish and Christian scriptures, Greek and Roman philosophy, Egyptian religion and Eastern mysteries could uneasily touch, merge and reconcile under the creative direction of poets, philosophers and priests.[4] These sapiential specialists included Plato, Orpheus and Pythagoras, the latter understood as historical figures by premodern humanists. Apuleius adds Psyche, hero of her own wisdom quest, and the votaries of Isis, the Hellenized Egyptian divinity who oversees the de-transformation of the golden ass. The term 'wisdom literature' was developed in the nineteenth century in order to group together a disparate set of texts from the Hebrew Bible that share features with neighbouring collections of proverbs and instruction as well as show some affinity with Greek philosophy and myth.[5] Although modern Bible scholars have criticized the category as ahistorical, the idea of wisdom literature nonetheless resonates with Renaissance humanists' interests in comparative religion, myth, poetry and philosophy.

Ancient wisdom writings are ripe with environmental implications: biblical wisdom literature often praises the sublimity and variety of creation, while pagan discourses from Orphism to Stoicism cultivated natural philosophy as spiritual exercise. Unhae Park Langis grasps wisdom's simultaneously cosmic and cultural reach in the idea of a *wisdom ecology*:

> A wisdom ecology embraces various facets of experiential life – numinous, cognitive, emotional, and ethical – and manages their constant imbrication and interaction in a world of thinking and living beings. The numinous development of wisdom involves intuition, contemplation, and ongoing spiritual practice. Wisdom in the world operates at the cognitive-affective-ethical interface as we practice protean skills in social and natural environments.[6]

A wisdom ecology comes into play wherever ideas saturate setting in a feedback loop of attention, care, and new thought that pays respect to nature as a thinking thing, world soul or cosmic organism. The wisdom perspective gathers *Midsummer*'s far-flung references to Greece, Thrace, Egypt, India and Ethiopia into a shared quest for the healing of body, soul and earth.

I begin at the end of the play, with Theseus' speech on the imagination, which describes the cosmic aspirations of the Orphic poet. I then turn to Isis/Titania and Lucius/Bottom as Apuleian/Ovidian representatives of wisdom's orbit. Reading from the sun-kissed edge of the play's fifth act to the yawning, yearning hee-haw at its nocturnal centre, I inventory the geographical sweep and environmental attentiveness of the play's wisdom work. Finally, I seek intimations of Isis in her role as both maritime protector and goddess of childbirth in Titania's elegy for the Indian votaress. Titania is a new Isis, a transnational wisdom muse of natural mutability and ethnic multiplicity who enters *Midsummer* from Ovid and Apuleius. Environmentalizing the two *Metamorphoses* also globalizes them, bringing forward their diverse locales, sources and counterparts and inviting them to cross-pollinate in Shakespeare's generative landscape of allusions and life-forms.[7] Across these three metamorphic and mythopoetic texts, the plasticity of souls in tropic environments composed of moving images and living things is also caught up in the migration of persons and ideas among disparate schools of wisdom.

Shakespeare's Orphic forest

In Act Five of *A Midsummer Night's Dream*, Theseus is planning the evening's entertainment. Among the rejected proposals is a play about Orpheus:

> Theseus: *[reads]* The riot of the tipsy bacchanals
> Tearing the Thracian singer in their rage.

That is an old device, and it was played
When I from Thebes came last a conqueror.

5.1.48–51[8]

In the *Metamorphoses*, the death of Orpheus concludes the Orphic sequence that begins in Book 10, with the death of Eurydice. When Orpheus returns alone from the land of the dead, he draws the forest and its sentient inhabitants around him through his music, calling the forest itself into being as a sensitive landscape that shadows lyric performance with memories of metamorphosis. As a mythic culture hero capable of taming wild beasts and summoning trees with his song, Orpheus orchestrates an originary instance of what Francesca Martelli calls the 'natureculture' of the *Metamorphoses*.[9] The 'virgin laurel [*innuba laurus*]' (*Met.* 10.92) who joins the Orphic parliament of trees bears her own tale of violence, change and poetic capacity in her shrinking bark and bending boughs: the Ovidian environment chiastically discovers nature and culture in each other. In the Orphic groves of the *Metamorphoses* and their Shakespearean reseeding in *Midsummer*, memory, mortality and aspiration infuse every instance of what is.

Returned from the dead in *Metamorphoses* 10, Orpheus recalls his grander efforts before taking up 'a gentler lyre' to sing of love and loss:

From Jove, O Muse, my mother, for all things come
From Jove, inspire my song! I have often sung
His power before, his wars against the giants,
His thunderbolts, but now the occasion seeks
A gentler lyre.

Met. 10.148–53; *trans. Humphries*[10]

Ovid's lyric Orpheus emerges as a response not only to the more tragic-heroic Orpheus of Virgil's *Georgics* but also to the sacerdotal Orpheus hypothesized by the philosophers and worshipped in Orphic circles.[11] Ovid at once privatizes the vocation of the poet by documenting his turn to a 'gentler lyre [*leviore lyra*]' and acknowledges the cosmic ambitions of the Orphic poet transmitted in hymns and theogonies.[12] Orpheus's mythopoetic inquiries into the origin of the cosmos are matched by Ovid's own story of the creation of the world in Book 1 and by Pythagoras' discourse in Book 15.[13] The existence of an Orphic 'religion' (the very term itself tainted by the relative lateness of Christianity as a phenomenon) is much debated by scholars, and many prefer to speak of Orphic arts or Orphic literature rather than of Orphism as an established cult.[14] Other scholars argue for the existence of Orphic rites in the period stretching from before Plato into the Imperial period.[15] The Orphic corpus reflects Near Eastern influences, including Hurrian-Hittite, Babylonian, Phoenician, Ugaritic and Hebrew creation stories.[16] The non-Hellenic character of Orpheus is marked by the poet's Thracian epithet and the orientalizing presentation of Dionysus, whose cult the early Orphics likely both followed and reformed.[17]

Bridging the expanse of the *Metamorphoses*, these Orphic legacies remix in *A Midsummer Night's Dream*. At the start of Act Five, just before the 'Thracian singer' reference, Theseus delivers his essay on the imagination:

> Hippolyta: 'Tis strange, my Theseus, that these lovers speak of.
> Theseus: More strange than true. I never may believe
> These antique fables, nor these fairy toys.
> Lovers and madmen have such seething brains,
> Such shaping fantasies, that apprehend
> More than cool reason ever comprehends.
> The lunatic, the lover, and the poet
> Are of imagination all compact.
> One sees more devils than vast hell can hold:
> That is the madman. The lover, all as frantic,
> Sees Helen's beauty in a brow of Egypt.
> The poet's eye, in a fine frenzy rolling,
> Doth glance from heaven to earth, from earth to heaven,
> And, as imagination bodies forth
> The forms of things unknown, the poet's pen
> Turns them to shapes, and gives to airy nothing
> A local habitation and a name.
> Such tricks hath strong imagination
> That if it would but apprehend some joy
> It comprehends some bringer of that joy;
> Or in the night, imagining some fear,
> How easy is a bush supposed a bear!
>
> 5.1.1–22

Scholars have long noted that while Theseus aims to champion 'cool reason' over poetic and erotic fantasy, Shakespeare himself seems to be formulating a defence of poetry.[18] The gesture of convening 'the lunatic, the lover, and the poet' in a shared imaginative compact reaches back to Plato, especially the *Symposium*, where the philosophy of eros delivered by the mystagogue Diotima via Socrates shares the banquet table with the mythopoetic (and quasi-metamorphic) musings of Aristophanes.[19] Erasmus's *The Praise of Folly*, a source for *Midsummer*'s festive philosophizing, cites Plato, who 'wrote, *that the passion and extreme rage of feruent louers was to be desired and embrased, as a thing aboue all others most blissful.*'[20] The lover and the philosopher transcend their ordinary perceptual apparatus through the ardent power of love, an ascent scaffolded by the handholds of poetic imagery. Such a vocation for poetry is broadly shared in Platonizing circles: Sufi poets, for example, drawing on Zoroastrian, Platonic, and Christian as well as Islamic ideas, describe the work of lyric as 'weaving a festive dress for the bride Meaning.'[21] In Ovidian fashion, every flower in Persian mysticism contributes to the erotic landscape: 'The narcissus is all eyes to look at the friend. The cypress reminds the poet of the

Beloved's graceful stature, the hyacinth of his dark curls.'[22] According to the Jewish philosopher and physician Leone Ebreo (1460–*c.* 1530), 'Fantasia [*l'immaginazione e fantasia*] arranges, distinguishes, and ponders sensory things, and discerns grace and beauty in many other sources, moving the soul to the delight of love.'[23] Ebreo's sources include the Hebrew Bible, Maimonides, and the Zohar; Arabic texts by Al-Farabi, Avicenna and Averroes; and works by Plato, Pico and Ficino. Ebreo also cites Ovid on the origins of the world and on mythological figures such as Hermaphroditus and Narcissus.[24]

The Christian humanist Erasmus, the Sufi mystics, and the Jewish philosopher Leone Ebreo all participate in a variegated wisdom ecology whose hierarchical yet interpenetrating levels of being irradiate the world with the hope of meaning.[25] This is not the "flat ontology" of the New Materialism but rather a laddered and layered organization of Being in which humans strive and struggle above the beasts and below the angels, yoking their animal senses to their divine *imago* in order to travel the heights and depths of the cosmos. Orpheus is an exemplary rhapsode of an experiential environment defined by descent and ascent, *katabasis* and *anabasis*. The poet whose eye 'in a fine frenzy rolling / Doth glance from heaven to earth, from earth to heaven' evokes the sojourn of Dante in the service of Beatrice and behind Dante, of Orpheus and Eurydice.[26] The 'fairy toys' and 'antique fables' grouped together by Theseus in his essay on the imagination had earlier converged in *Sir Orfeo*, a Middle English lay in which Queen Heurodis (Eurydice) is stolen by fairies and taken not to the underworld but to Fairyland.

According to the neo-Platonist Proclus, 'All theology among the Greeks is sprung from the mystical doctrine of Orpheus.'[27] Pythagoras is one of these Orphic students; he was familiar enough to Elizabethans in the 1590s that a play, now lost, was written about him in 1596, around the same time as *Midsummer*.[28] For Ovid's Pythagoras, the soul or *anima* organizes reason, passion, memory, perception and imagination, granting provisional identity to those who suffer the ecstasy of change in a landscape alive with potentiality. In the words of Giulia Sissa, both Ovidian metamorphosis and Pythagorean metempsychosis yield a 'productive becoming', a life that is 'frozen, stretched, displaced, relocated. Nothing is purely and simply annihilated, nobody dies entirely.'[29] Lynn Enterline notes the Ovidian flow between the poet's creative *animus* (mind) in the first line of the *Metamorphoses* and the *anima* or soul; both words are identified with voice, breath, and breeze, like the Stoic *pneuma*, the Hebrew *ruach* and the Sanskrit *prana*.[30] Threading *anima* through the image-making capacities of many kinds of being, Leone Ebreo attributes imagination to any creature with a 'sensitive soul', a common faculty that joins in friendship 'all the species of terrestrial, aquatic and winged animals.'[31] A menagerie of animated images flickers through Theseus' wry *conclusio*: 'in the night, imagining some fear,/ How easy is a bush supposed a bear!' In Theseus' shadow theater of ambient anxiety, ursine Callisto hides in a stand of shuddering Daphne's. In *Midsummer*'s Orphic forest, linguistic and creaturely orders of experience continually exchange attributes in a passionate pictorialism on whose nocturnal screen 'quick bright things come to confusion' (1.1.149).

The calmed sea is a mirror for shooting stars in the mermaid scene recollected by Oberon, possibly based on an actual courtly entertainment featuring fireworks and an artificial lake:

> Thou rememb'rest
> Since once I sat upon a promontory
> And heard a mermaid on a dolphin's back
> Uttering such dulcet and harmonious breath
> That the rude sea grew civil at her song
> And certain stars shot madly from their spheres
> To hear the sea-maid's music?
>
> 2.1.148–54

This civil siren blends Arion and Orpheus in a uniquely feminine portrait of poetic capacity. The mermaid's 'dulcet and harmonious breath' evokes Pythagorean cosmo-musicology, in which the music of the spheres was tuned by celestial beings identified as sirens, muses, or angels.[32] The mermaid's biform body and dolphin partnership harmonize her apportioned being to the channels of the universe.[33] As Hugh Grady argues, the scene's refulgent utopianism both glorifies the Elizabethan state and suggests dynamic worlds of understanding and experience beyond the purview of power.[34]

When Theseus disparages the lover for seeing 'Helen's beauty in a brow of Egypt', his casual racism also inadvertently acknowledges the historic flow between Greek and North African philosophy: Neoplatonism was born in Egypt, and the mythic-historic Orpheus and Pythagoras were said to have studied there.[35] Choosing fair Helena over dark Hermia, Lysander disparages his former love as a 'tawny Tartar' and an 'Ethiope' (3.2.257, 263), calling attention to the permeability of aesthetic canons that would fascinate Shakespeare throughout his career, from the sonnets to the Dark Lady to that consummate 'brow of Egypt' Cleopatra, whom Antony calls 'this great fairy' (4.9.12). What is true of beauty is also true of wisdom. William Baldwin begins his *Treatise of Morall Philosophie* with a global mapping of philosophy's origins:

> There is great diversitie among Writers, some attributing it to one, and some to another: as the Thracians to Orpheus, the Grecians to Linus, the Libians to Atlas, the Phenecians to Ocechus, the Persians to their Magos, the Assyrians to their Chaldees, the Indians to their Gimnosophistes, of which Budas was chiefe, the Italians to Pithagoras, and the French-men to their Druides.[36]

A lineage stretching from Orpheus to Pythagoras winds through Thrace, Assyria, and India as well as Greece and Italy, locales touched upon in *Midsummer*. Both Titania and Oberon have visited India and may be Indian themselves.[37] The story of Pyramus and Thisbe begins in Babylon and ends in Nineveh, contributing to what one scholar calls Ovid's Babylonika.[38] In the woods, Flute/Thisbe addresses Bottom/Pyramus as 'Most bristly juvenile, and eke most lovely Jew' (3.1.89), pointing to the story's Near Eastern setting. The Sophia of the Septuagint is a 'lovely Jew': 'Wisdom is more moving than any motion: she passeth and goeth through all things.'[39] Orpheus's tale of Myrrha ends 'in Saba land [*Sabea*]' or Sheba, identified with Ethiopia in the Geneva Bible and with the Shulamite or beloved of Solomon in the sapiential Song of Songs.[40] The heaven and earth scanned by the Orphic poet shelters both a global geography and a wisdom ecology, a world alive with multiple traditions and ensouled by manifold intelligences. Pythagorean metempsychosis is a thought experiment inviting the philosopher to consider their kinship with other living beings in different times and places.[41] The *sophia* that emerges in the dense roots, sheltering arborage, and vaulting heavens of *Midsummer* is Hellenic, Roman, Egyptian, Persian, Hebrew and Indian, and the *anima* she cultivates manages a set of creative capacities shared among many kinds of being linked through love and friendship. It is this Sophia that stares longingly back at Bottom from the depths of his dream.

Bottom translated, Titania descending: a creaturely theophany

Empedocles, a follower of Pythagoras, was supposed to have recalled his own soul's journey as 'boy and girl, bush, bird, and a mute fish in the sea'.[42] Bottom is *Midsummer*'s hairy Pythagoras and goofy Empedocles. Although there is much Ovid in the shaping of Bottom, another *Metamorphoses*, *The Golden Ass* of Apuleius, is the more immediate analogue. The 1566 English translation of *The Golden Ass* emphasizes the text's transnational character:

> Lucius Apuleius African, an excellent follower of Plato his sect, borne in Madaura, a country sometime inhabited by the Romans and under the jurisdiction of Syphax, situate, and lying upon the borders of Numidia & Gaetulia, whereby he calleth himself, half a Numidian, and half a Gaetulian: and Sidonius named him the Platonian Madaurdence. His father, called Theseus, had passed all offices of dignity in his country, with much honour: his mother named Salvia, was of such excellent vertue, that she passed all the dames of her time, born of an ancient house, and descended from the noble Philosopher Plutarch, and Sextus his Nephew.[43]

Scholars in recent decades have reclaimed Apuleius as a multi-cultural author whose 'Romano-African identity' linked him to Punic Carthage as well as Greece, Rome, Egypt and Mauritania.[44] *The Golden Ass*, Apuleius tells his reader, was composed on 'Egyptian paper written with a ready pen of Nile reeds'.[45] Richard Fletcher styles him as an "Afro-Platonist," a moniker derived from Augustine's characterization of his countryman as both *Afer* and *Platonicus*. Identifying philosophy and the soul with migration, hybridity and exile, Apuleius' Afro-Platonism 'is powerful and performative' in its engagement with Greek, Roman and Persian texts and exemplars.[46] Through his Orphic and Pythagorean progenitors, Egyptian teachers, Alexandrian systematizers and Abrahamic exegetes, Plato becomes not the Hellenic advocate of pure contemplation but the global rhapsode of an eros that saturates the world with order, sentience and meaning.[47]

At the end of *The Golden Ass*, Isis rises from the waters like Venus Anadyomene. Resplendent in moonlight and seafoam, Isis presents herself as the distillation of multiple goddesses:

> Behold Lucius I am come, thy weeping and prayers hath moved me to succor thee, I am she that is the natural mother of all things, mistress and governess of all the Elements, the initial progeny of worlds, chief of the powers divine, Queen of heaven, the principal of the Gods celestial, the light of the Goddesses, at my will the Planets of the air, the wholesome winds of the Seas, and the silences of Hell be disposed, my name, my divinity, is adored throughout all the world, in divers manners, in variable customs, and in many names, for the Phrygians call me the mother of the Gods: The Athenians, Minerva: the Cyprians, Venus: the Candians, Diana: the Sicilians, Proserpina: the Eleusians, Ceres: some Juno, other Bellona, other Hecate: and principally the Ethiopians which dwell in the Orient, and the Egyptians which are excellent in all kind of ancient doctrine, and by their proper ceremonies accustom to worship me, do call me Queen Isis.
>
> Ap. *Met.* 11.47

Appearing 'in divers manners, in variable customs, and in many names', *polyonymous* and even *myrionymous* Isis receives her proper habitation and name in Ethiopia and Egypt: 'Queen Isis'.[48] Discovering the one in the many by progressively unpeeling her accumulated names and origins to identify herself with Ethiopian and Egyptian wisdom, Isis embodies the Afro-Platonism of Apuleius.

Apuleius and his predecessor Plutarch, also a Platonist, identified Isis with nature and thus with generation, mutability, decay, and renewal. Paralleling similar descriptions of Woman Wisdom in Hebrew wisdom literature and the Septuagint, Plutarch identifies Isis with 'the female principle of Nature' in her mode of becoming: 'she turns herself into this or that thing and is receptive to all manner of shapes and forms'.[49] In *The Golden Ass*, Isis appears to Lucius in 'a vestiment … of fine silk yielding divers colours, sometime yellow, sometime rosie, sometime flamy, and sometime (which troubled my spirit sore) dark and obscure' (XI.47). The multi-coloured undergarment represents the mutability of nature as well as the geographical dispersion of the goddess herself. The black cloak wound around this gown evokes not only the nighttime sky but also the urge towards interpretation that Isis solicits in her mode of mourning. Isis wore black when she gathered the scattered body parts of Osiris, an act that became an emblem for the hermeneutic projects of the Renaissance.

Associating Orpheus and Isis, Pierre Hadot reconstructs a line of natural philosophy that takes the veil of the goddess as the emblem of a science of sympathies that runs counter to Promethean technology. In this tradition, 'the work of art, the discourse or poem, is a means of knowing Nature … for the artist espouses Nature's creative movement, and the event of the birth of a work of art is ultimately a mere moment in the event of the birth of Nature.'[50] Shakespeare, Ovid, and Apuleius participate in this poetic-immersive understanding of nature's poiesis, which they rehearse and recreate through metamorphosis.

Isis enters *A Midsummer Night's Dream* only obliquely, carried on the back of Apuleius' golden ass like an errant Europa. As James McPeek points out, 'governess' and 'progeny', unusual words featured in Adlington's translation of the aretalogy of Isis; show up in

Titania's long speech about climate disruption. She calls the moon 'the governess of floods' (2.1.103) and claims that 'this same progeny of evils' (weather inversions and disturbances) 'comes from our debate, from our dissension' (2.1.115-16).[51] What is orderly in the speech of Isis has come undone in the speech of Titania, and what is singular in Apuleius becomes relational in Shakespeare. Both deities manage a dynamic cosmos composed of interdependent systems of weather, agriculture, and cult. With Apuleius already on tap for Bottom's transformation, Shakespeare was likely recollecting the theophany of Isis when he composed Titania's longest, and most explicitly environmental, speech: both are moon, weather, and fertility goddesses who draw multiple feminine divinities into their sublime aura, and both appear to ass-men in sensitive landscapes inhabited by mobile, malleable souls.

In Ovid's *Metamorphoses*, Io, transformed into a white heifer to protect her from the wrath of Juno, ultimately regains her former shape in Egypt, where she is worshipped as a divinity: 'Now, as a Goddess, is she had in honour everywhere / Among the folk that dwell by Nile yclad in linen weed' (*Met.* 1.939–40; Golding trans.). The deity in question is Isis, who was linked to Io as early as the sixth century BCE; the story of Io is depicted on the walls of the Temple of Isis at Pompeii. The *translatio* of Io into Isis contributed first to the Hellenizing and later to the Romanizing of Egypt after Actium.[52] Isis appears again in *Metamorphoses* 9 to oversee the transformation of Iphis from female to male (*Met.* 9.666–797), a transgender story that is taken up by John Lyly in his *Gallathea* (1584?) and by Shakespeare in *Twelfth Night*.[53] Ovid also addresses a poem to Isis as the goddess of childbirth in his elegy on Corinna's abortion and its dangerous aftermath (*Amores* 2.13).[54] In Ovid, Isis represents a Romanized Egyptian cult associated with animal metamorphosis, female worship, and the dangers of pregnancy in a wisdom ecology that courses among diverse cults, genders and species.

Ovid is everywhere in *The Golden Ass*: Olga L. Levinskaya and M. Nikolsky have argued that the whole novel, including Lucius' wanderings and the trials of Psyche in the inset tale, are modelled on the Io episode of the *Metamorphoses*.[55] And Ovid resounds throughout the book: the episode leading up to Lucius' momentous transformation, for example, is framed by a virtuoso ekphrasis of an Actaeon and Diana sculpture group at the entry of the house of Byrrhena (2.4) that blends the landscapes of forbidden theophany in the two works into a joint reflection on art, nature, and divinity.[56]

In Apuleius, the dream vision of Isis initiates Lucius' transformation back into a human being, like Io becoming Isis on the banks of the Nile. Shakespeare breaks the theophany into two parts: Titania's first encounter with Bottom and Bottom's recollection of his dream the following day. Shakespeare's intentions are largely comic, but traces of the sublimity of Apuleius's 'Isis book' linger, as Jan Kott and others have argued.[57] In *A Midsummer Night's Dream*, Diana's secret grotto becomes the clearing in the forest where the Rude Mechanicals rehearse their Ovidian play. But whereas Actaeon's sylvan encounter triggers Diana's rage, in *Midsummer* Titania is smitten by the hybrid creature whose simple song awakens her from her drugged slumber:

Bottom: I see their knavery. This is to make an ass of me, to fright me, if they could. But I will not stir from this place, do what they can. I will walk up and down here, and I will sing, that they shall hear I am not afraid.

[He sings]

The ousel cock so black of hue,
With orange-tawny bill;
The throstle with his note so true,
The wren with little quill.

Titania: *[awaking]* What angel wakes me from my flow'ry bed?

Bottom: *[sings]* The finch, the sparrow, and the lark,

The plainsong cuckoo grey,
Whose note full many a man doth mark,
And dares not answer 'Nay' –

for indeed, who would set his wit to so foolish a bird? Who would give a bird the lie, though he cry 'Cuckoo' never so?

Titania: I pray thee, gentle mortal, sing again.
Mine ear is much enamoured of thy note.
So is mine eye enthrallèd to thy shape;
And thy fair virtue's force perforce doth move me
On the first view to say, to swear, I love thee.

3.1.116–34

For lyrical Helena, bird song evokes Orphic harmonies, 'your tongue's sweet air / More tuneable than lark to shepherd's ear' (1.1.193). Bottom's rough music is more Pan than Apollo, yet the inventory of birds recalls a whole genre of wisdom writing that includes Chaucer's *Parlement of Fowls* (1382?) and Attar's Sufi *Conference of the Birds* (1171).[58] Bottom uses the song's simple rhythms to reorient his altered state to the sylvan environment, walking 'up and down' like a small child or animal pacing to calm itself in an unknown situation. Bottom's plainsong is an interface, a tuning tool and a therapy.

In Apuleius, the goddess appears to the creature, who gratefully receives her instruction. In Shakespeare, the creature is revealed to the goddess, who intuits 'fair virtue's force' in his voice and shape. The eye and ear of the goddess become erotic portals for a stream of impressions that attach to Bottom insofar as he channels the thicket teeming around them. According to one line of thought in the history of religions, what distinguished the divinities of the mystery cults (including Isis, Orpheus, and Dionysus) was their own mortality, their own participation in the cycle of birth, death, and renewal.[59] In *Midsummer*, the goddess has descended: in having creaturely life reveal itself to her,

Titania partakes in the mystery she administers. Rather than the tragic sparagmos of Orpheus, however, this is a comedy of the phallus, its ridiculous donkey hugeness displaced onto Bottom's long hairy ears. Titania drifts to the forest floor by passing through a moiré of Apuleian paradigms.

Titania-as-Isis morphs into Psyche, the beauty who loves a beast, and then devolves into the Corinthian matron enamored of the ass in Book 10. Apuleius compares the Corinthian woman to Pasiphae (10.46), casting Lucius as a comic minotaur (with Theseus/Oberon waiting in the wings).[60] In both Apuleius and Shakespeare there are also echoes of Europa decking the jovial bull with garlands (*Met.* 2.833–75). Insofar as the affair has been orchestrated by Oberon as a humiliation, Titania descends even lower, into the murderess who condemned to suffer ritual debasement by being publicly submitted to sex with the Lucius-ass before being torn apart by wild animals (10.46).

In a double translation that 'passes through animality', both players are initiates into the mysteries of 'fair virtue's force'.[61] The phrase handles virtue in a broadly vitalist sense, to indicate the innate potential (*dynamis*) and actualized energy (*energeia*) of beings and things in the world. In this ancient extra-moral and environmental understanding of virtue, human excellence is always 'set in relation to physical powers that worked, for good or ill, as part of a broader ethical ecology'.[62] Ovid's etiologies probe virtue in this material and organic sense, as Vin Nardizzi argues in his study of Gerard's *Herball* of 1597/98.[63] In *Midsummer's* bottom-heavy theophany, 'fair virtue's force' flows between the man-beast and the goddess-descended in an ecological exchange of sonorous rhythms, forest foods, and animate breath that together compose an Ovidian-Apuleian biosphere.

The theophany is completed in Shakespeare's play when Bottom wakes up. Kott locates Bottom's allusion to St Paul in the context of Renaissance Platonism, where Paul takes his place alongside Orpheus, Moses, Plato and Apuleius.[64] Bottom's garbled parody of St Paul's letter to the Corinthians cleverly recalls Lucius' own return to Corinth and to human form (4.1.205–10; 1 Cor. 2:9–10).[65] Behind Paul lie other wisdom texts. Leone Ebreo, praising the imagination, cites Ecclesiastes: 'Solomon says that the eye is not satisfied with seeing, nor the ear filled with hearing [Eccl. 1:8].'[66] The donkey sees with his "marvelous hairy ears" and his moist, quivering muzzle, his porous orifices yielding the synesthesia of the mystics alongside the knowledge of 'sweet hay' (4.1.24, 43).[67] Now it is Bottom's ears and eyes that surge beyond their normative capacities in order to grasp the sound of the world. Speaking in broken paradoxes that invite the real to shine through, our oafish Orpheus is composing his own wisdom poem to creation.

Festivals of Isis: a ship and an abortion

In *The Golden Ass*, Isis appears to Lucius rising from the sea, a posture from the pictorial tradition of the *Aphrodite Anodyomene* that culminates in Botticelli's *Birth of Venus* and also animates Plutarch's and Shakespeare's descriptions of Cleopatra at Cydnus. This tradition was syncretized in representations of *Isis Pelagia* or *Isis Velificans,* which depicted the goddess holding a sail, its floating form often mirrored in a cloak billowing behind her.[68] The cults associated with these marine goddesses attempted to manage the risks of seafaring through propitiatory rites that acknowledged the roiling waters as sites of cosmic flux and creativity. Aphrodite Anadyomene is also Venus Genetrix, the mother of all living things, as explored in the Venus passages of Lucretius's *De Rerum Natura*, a text that impacted both Ovid and Apuleius.[69]

Isis Pelagia was honored in the *navigium Isidis* or maritime festival of Isis, which Lucius observes on the morning after his theophany. In these rites, a model ship was filled with votive offerings and then sent down the shoreline to Isis's temple:

> The great Priest compassed about with divers pictures according to the fashion of the Egyptians, did dedicate and consecrate with certain prayers a fair ship made very cunningly, and purified the same with a torch, an egg, and sulfur, the sail was of white linen cloth, whereon was written certain letters, which testified the navigation to be prosperous, the mast was of a great length made of a pine tree, round, and very excellent with a shining top, the cabin was covered over with coverings of gold, and all the ship was made of Citron tree very fair, then all the people as well religious as prophane took a great number of fans replenished with odors and pleasant smells, and threw them into the sea mingled with milk, until the ship was filled up with large gifts and prosperous devotions.
>
> Ap. *Met.* 11.47

The *navigium Isidis* featured festival boats filled with offerings associated with the dismembered body of Osiris. Votaries would 'beat their breasts and imitate the sorrows of an unhappy mother' in rites that resembled mystery plays.[70] I suggest that we read this passage in relation to Titania's recollection of her Indian votaress:

> Set your heart at rest.
> The fairyland buys not the child of me.
> His mother was a vot'ress of my order,
> And in the spicèd Indian air by night Full
> often hath she gossiped by my side,
> And sat with me on Neptune's yellow sands,
> Marking th'embarkèd traders on the flood,
> When we have laughed to see the sails conceive
> And grow big-bellied with the wanton wind,
> Which she with pretty and with swimming gait
> Following, her womb then rich with my young squire,
> Would imitate, and sail upon the land
> To fetch me trifles, and return again
> As from a voyage, rich with merchandise.
> But she, being mortal, of that boy did die;

And for her sake do I rear up her boy;
And for her sake I will not part with him.

2.1.121–37

The idyll has been interpreted in the context of English trade with India, and Walter Staton associates the pregnancy and death of the votaress's with Ovid's Calisto.[71] Possible connections to the *navigium Isidis* have not been noted, as far as I am aware. The metaphoric mirroring of woman and ship on sand and sea miniaturizes the 'big-belllied' sailing vessels into toy boats or festival floats, the swell of her moving garment echoing the vessels at sea. The votaress 'sails upon the land' laden with 'trifles' like a carnival wagon; becoming an image of movement and transport, her flowing, fulsome person merges modalities of being in the manner of Oberon's Orphic mermaid. The lyric dance of images, anticipating the 'pageants of the sea' in *The Merchant of Venice* (1.1.11), merges the votaress *velificans*, the gifts she brings to Titania, the ships filled with luxury goods, and the golden seascape in a memory theater that stages friendship as festival, sovereignty as solidarity, and coastline as concourse of souls. Although Shakespeare is unlikely to have had this passage directly in mind in *Midsummer*, Apuleius' description echoes elements of Cleopatra's barge at Cydnus, a political-theological spectacle that channeled the marine iconography shared by Venus and Isis. Veils become sails as the images bob and jostle in an ocean of myth and meaning, navigating the human enterprises, primal generativity, sensory abundance, and risks of life by and on the sea.[72]

The real danger here, however, is not shipwreck but childbirth. Both Diana and Isis presided over pregnancy, as Ovid dramatizes in his elegy to Isis (*Amores* 2.13). The speaker contrasts the *onus* of pregnancy (load, weight, burden) with the *munera* (votive gifts) of petitionary prayer in order to plea for Corinna's survival following her abortion.[73] Titania's tale, cresting in maternal mortality, harbors similar resonances in words like 'merchandise' and 'trifle', which simultaneously figure the ship's cargo, the unborn child, and gifts for the fairy queen. Both Shakespeare's idyl and Ovid's poem recall female folk practices around the dangers of pregnancy and the claiming of miscarried, aborted, orphaned and deceased children by divine or daemonic feminine forces.[74] In the Middle English *Sir Orfeo*, women who die in childbirth are taken to Fairyland, not an underworld so much as a holding ground for those who have suffered certain kinds of deaths.[75] Are the 'trifles' fetched by the votaress *munera* designed to immunize their giver against the dangers of childbirth? Has a 'fairy toy' (5.1.3), a cunning effigy memorializing a traumatic pregnancy, been left in the place of the Indian boy?

The Indian votaress belongs to a female worship community, like the cults dedicated to Isis in Egypt, Greece, Rome and the Hellenized world. Ovid's Corinna is also a votaress: Ovid reminds Isis that his girlfriend has faithfully sat 'in ministration to thee on the days fixed for thy service [*saepe tibi sedit certis operata diebus*].'[76] The frescoes in the Temple of Isis preserved at Pompeii represent Roman women, slaves, and freedpersons of several ethnicities worshipping Isis together; in *Midsummer*, the worship of Titania crosses cultures in search of a sophianic female community that is attuned to the affordances of locale yet also restlessly migratory and translational. The sea as primal source of generation dissolves the twin themes of pregnancy and seafaring in a liquid matrix of natureculture and lifedeath. At once intimately dyadic and environmentally plentiful, her elegy for the Indian votaress is a kind of Isaic wisdom poem, recollecting bits of worship and refiguring them in a seascape that freely mixes Corinthian/Graeco-Roman and Indian/Egyptian elements. The speech itself becomes a *navigium Isidis,* a lyric hymn pregnant with *munera* that both mourn and enshrine feminine friendship, service, and grace within the onward march of imperial power and its translation projects.

Passing through a rich wardrobe of Ovidian and Apuleian guises, Titania gleams with the breadth of creation, understood as both biodiversity and cultic diversity. Like Io and Psyche, she suffers love and loss by descending into mortal existence. Titania belongs to a web of Isis-meditations that float through Shakespeare's work on Roman, Greek and Egyptian wings. The *navigium* of images launched in *Midsummer* comes ashore in the Cydnus of *Antony and Cleopatra*, Imogen's studiolo in *Cymbeline*, and the Ephesus of *Pericles*.[77] In these and other works, Shakespeare participates in a renaissance of interest in Isis as a Hellenized and Romanized goddess of cosmic connection with roots and branches in North Africa and the Middle East. According to Peggy Muñoz Simmonds, Isis symbolizes for Shakespeare and his contemporaries 'the central Neo-Platonic idea of the "Many in the One," an image of mystical reunification after dismemberment that is essential to the Orphic ritual hidden within all true Renaissance tragicomedies.'[78] Unlike Apuleius, Ovid was no Platonist, but his stories blended readily with Platonic mythopoesis and its marriage with Hellenized Egyptian wisdom. In the judgment of Stuart Sillars, *Midsummer*'s sojourning with Ovid and Apuleius as well as Erasmus and St Paul yields a work 'more complex and transnational ... than often considered'.[79] François Laroque concurs: 'Shakespeare's poetry defines itself as a geography of desire, a mapping out of knowledge through fantasy across social, sexual, and racial differences,' to which we can add variations and modulations among species and life forms.[80] Shakespeare learns from the Orpheus and Isis of Ovid and Apuleius how to merge the *onus* of loss and travail and the *munera* of votive gifts into an 'impure aesthetics' that beckons beyond the ceaseless fungibility of the commodity in order to tune into cosmic melodies and animal sounds.[81] He learns, that is, to listen to 'fair virtue's force', a potentiality at once poetic and spiritual, psychological and environmental, that takes sustenance from the shape-shifting, gender-tenderized, Sophia-seeking anima of Ovid and Apuleius.

At once repository and repertoire, the wisdom element running through Shakespeare's plays taps Hebrew, Graeco-Roman and folk traditions and brushes against their sources and counterparts in other parts of the world, opening the play to an astonishing range of creaturely tonalities, from the donkey's plaintive bray to the dolphin-mermaid's dulcet harmonies. *Midsummer*'s transnational wisdom irradiates the play's environmental sensibilities: the two resound together, like lark to shepherd's ear. Sophia dances in a global setting of intellectual exchanges that also orchestrate a suite of sympathetic concordances among sentient beings. Shakespeare's metamorphic forest is a luminous and variegated philosophical, aesthetic and natural-historical space, a Fairyland whose rules of conservation and non-contradiction run counter to the instrumentalizing and otherizing narratives of imperialism, capitalism and racism. In *Midsummer*, Shakespeare approaches wisdom literature as a seed bank, a precious compendium of manifold forms of life stored in proverb, image, myth and hymn, ready to burst into new bloom under conditions of attention and care.[82]

Notes

1. Works of note on Shakespeare and Ovid include Barkan 1986; Bate 1993; LaFont 2014; and Starks 2019.
2. E.g. Starnes 1945; Kott 1987: 29–68; Generosa 1945.
3. Kott 1987. See also Kott 1964, a landmark essay.
4. Wind 1968.
5. Kynes 2021.
6. Langis and Lupton forthcoming.
7. Martelli 2020: 79–81.
8. All citations from Shakespeare are from Taylor et al. 2016.
9. Martelli 2020: 39.
10. For the *Metamorphoses*, I am using the translation of Humphries (1955), and the Arthur Golding translation of 1567, for which see Nims 2000. All Latin citations are from Loebclassics.com.
11. Segal 1989: 54–72.
12. Meisner 2018: 159–60.
13. On the programmatic articulation of the *Metamorphoses*, see Myers 1994: 161–5.
14. Linford 1941; West 1983.
15. Guthrie 1993; Kingsley 1995: 117–20; Burkert 1985: 430–6.
16. Meisner 2018: 18–19.
17. Guthrie 1993: 39–41; Burkert 1985: 430–6.
18. E.g. Dent 1964.
19. Evans 2006.
20. Erasmus 1549: 153.
21. Schimmel 1992: 41.
22. Schimmel 2001: 77.
23. See Ebreo 2009: location 4568.
24. Ebreo 2009: locations 1871, 2937, 6450.
25. On the ecological dimensions of Erasmus, grounded in the Stoic concept of *oikeosis*, see Dealy 2017.
26. Clay 2014: 176.
27. Proclus, cited in Wind 1968: 58.
28. Borlik 2016.
29. Sissa 2019: 164.
30. Enterline 2000: 51.

31. Ebreo 2009: location 2049.

32. Lindley 2005: 14–18.

33. Compare Cleopatra on Antony: 'His delights / Were dolphin-like; they showed his back above / The element they lived in,' 5.2.87–8.

34. Grady 2008: 285.

35. Burkert 2004: 71–98.

36. Baldwin 1546: i.

37. Hendricks 1996. On Orpheus and India, see Adluri and Bagchee 2012.

38. Holzberg 1988.

39. Wisdom of Solomon 7:24 (KJV).

40. *Met.* 10.550, trans. Golding. 1 Kings 10.1, Geneva Bible note 'Josephus saith, that she was Queen of Ethiopia, and that Sheba was the name of the chief city of Meroe, which is an island of Nile.'

41. Cf. Sherman 2016.

42. Empedocles fr. B 117 in Wright 1995. Lloyd 1699: ix recounts the story.

43. 'Life of Lucius Apuleius, Briefly Described,' in Apuleius 1915.

44. Finkelpearl 2014: 1.

45. Ap. *Met.* 1.1.

46. Fletcher 2014.

47. Langis 2022 calls attention to the Alexandrian origins of Neo-Platonism and its subsequent development in Morocco, Syria, Arabia and Persia, flowering in Sufism. See also Schimmel 1997.

48. On many-named and infinite-named Isis, see Bricault 2020: 4.

49. Plutarch, 'Of Isis and Osiris,' Chapter 53, in *Moralia*. On Woman Wisdom in Hebrew scripture and the Septuagint, including her Egyptian and Canaanite analogues, see Perdue 2007.

50. Hadot 2006: 156.

51. McPeek 1972. 'Governess' appears only twice in Shakespeare; 'progeny' shows up five times, usually in the context of kinship. This is the only place where it appears in the context of natural processes.

52. Chance 2013; Bricault 2018.

53. Chess 2015.

54. Heyob 1975: 73.

55. Levinskaya and Nikolsky 2017.

56. *Golden Ass*, 2.4. See Harrison 2014: 96–7.

57. Kott 1987: 29–68; Barkan 1986: 260.

58. For a comparative study of these two works, see Baeten 2020.

59. E.g. Alvar 2008. The mystery cults as described by James George Frazer in *The Golden Bough* (1889–90) were later critiqued as a back-formation of Christianity (Meisner 2018: 238–9). Walter Burkert 1983 returned to cults of the dying god in the context of new archaeological and epigraphic materials as well as the literary record.

60. Lamb 1979: 478–91.

61. Kott 1974: 218.

62. Crocker 2019: 5.
63. Nardizzi 2019.
64. Kott 1974: 33–5.
65. The allusion has been much discussed: cf. Waldron 2012.
66. Ebreo 2009: location 1361.
67. Cf. Merrifield 2009: 84.
68. Bricault 2020: 100, 89, 95, 124. For the medieval and Renaissance tradition, see Heckscher 1956.
69. Asmis 1982. On Ovid and Lucretius, see Myers 1994.
70. Heyob 1975: 55.
71. Hendricks 1996: 48–51; Staton 1963.
72. This is the purview of the blue humanities. On which, see Mentz 2021, 2009.
73. *Onus* appears twice, *Am.* 2. 13.1 and 20; *munera* appears three times, *Am.* 2. 13. 22, 24 and 26.
74. Lamb 2000.
75. Allen 1964: 105.
76. *Amores II* XIII.17, trans. Christopher Marlowe.
77. On Isis in *Cymbeline*, see Simmonds 1992: 95–101.
78. Simmonds 1992: 100.
79. Sillars 2013: 118.
80. Laroque 2014: 70.
81. Grady 2008.
82. On wisdom as seeds, see Seneca, *Epistles* 38.2.

CHAPTER 8
METAMORPHOSIS IN A DEEPER WORLD
Claudia Zatta

Arne Naess is the founding father of deep ecology, a philosophy and social movement that asserts the intrinsic value of nature along with its beings, assigns an ontological function to relations, and reframes the status that human beings hold in the environment.[1] No longer a detached observer, the human being now participates in the natural world that surrounds her: embedded in a network of relations, her sense perceptions and emotional responses become constitutive of the beings she relates to. The result is the emergence of a natural world deeply interconnected where humans (and all the other beings) constitute integral parts of the whole. Naess contrasts this conception of the natural world and related experience with Einstein's. He notes that the theoretical physicist saw the external world as an impersonal arena and that studying it freed him 'from an existence dominated by wishes, hopes and primitive feelings'. By contemplating that world and discovering its ineffable laws, Einstein found 'inner freedom and security' and was able to silence his personal demands vis-à-vis the huge, human-independent reality he tried to understand.[2] For Naess instead,

> The way of liberation through 'natural history' is different: very little abstract thinking, very much seeing, listening, hearing, touching. The secondary and especially the tertiary qualities are in focus – the world of concrete contents – not the primary qualities studied in physics.[3]

The natural world also provides liberation, but in a different way: rather than breaking the chains that confine the self by means of abstract thinking, Naess' natural historian dissolves them by means of the senses, by seeing, hearing, and touching what is around himself by listening to its silent presence. An infinite diversified spectacle lies open to observation – creatures large and small, 'minerals, rocks, rivers and tiny rivulets, ... plants, animals small or big ones, plant and animal societies, tiny and great ecosystems.'[4] Even the most minuscule living being possesses an inherent meaningfulness, which it conveys not by words but by its very existence, by its colours and shapes, its textures and its multiple interactions with its own outside world, however small. No natural being exists in isolation: every being depends on other creatures and is part of a complex web of relationships. As for communication, it is a subtler affair than human language suggests, and the natural historian, attuned to the secret languages of the beings around her, listens to them.[5]

Ovid's *Metamorphoses*: an ontology of the detail

One might wonder whether Ovid too found 'liberation' while imagining the world of the *Metamorphoses*. His exile in Tomis on the Black Sea in 8 CE, immediately after the poem's first publication, points to an undercurrent of tension from which he may well have found relief as he was composing this masterpiece. Writing stories of metamorphosis could have been his form of escape. What is certain, however, is that Ovid's poem reveals an approach to the natural world similar to Naess'. He too was alert to the myriad beings that exist in the natural world – from rivers and trees to animals and insects to ponds, mountains and flowers. Likewise, he was aware of their meaningfulness, but one that artfully coalesced in the imagined landscape of his poem. For in *Metamorphoses* Ovid does not merely observe and describe the existence of natural beings *in* a system of relations; rather, he looks at their emergence *from* (a system of) relations. He tells us how the first laurel tree came into life, how the wolf started to howl in the fields of Arcadia, or how a row of poplars began to populate the banks of the Eridanus in northern Italy; he accounts for the surging of the spring Arethusa on the island of Ortygia, and the flowing of the river Acis in Sicily – and he conceives of all these and other emergences as the natural transformative outcomes[6] of human beings, *due* to relations. Indeed, rather than living in isolation, all his characters are cast in complex relational dimensions (erotic, social, and familial) involving unrequited lovers and those who have spurned them, hosts and guests, daughters or sons and fathers, husbands and wives, mothers and children, siblings or rivals, human beings and gods. At times these categories intersect and are mobilized in the same account, at others they do not. Still, they always present a relational entanglement that leads the human characters to their own transformations and often continues to affect, albeit in a new fashion, their existences once the transformations have occurred. In *Metamorphoses* the relational and often problematic nature of existence is thus a key to understanding the characters and their stories including the metamorphic outcomes. For even in their existences post-metamorphosis Ovid's characters find themselves interconnected, but in a larger scheme that transcends the mere human (and the divine) and involves the biosphere.

Significantly, Ovid's narratives of transformations align him in yet another way with Naess' approach to the natural world. In a foundational essay titled 'The World of Concrete Contents',[7] Naess offers philosophical support for the reflections that initiate this study, affirming that the ontology of the world we live in, the *Lebenswelt*, encompasses the perceptual and affective reception of the living being experiencing it. He writes,

> The most interesting interpretation of 'matter,' as far as I can see, is such that it comprises all that man even can experience in any state. And that the possibility is not excluded that other sensitive beings can experience additional 'things' which humans cannot . . . What we feel about something belongs to the world as we know it. What does not have such qualities is abstract structure. Environmentalists talk about reality as it is in fact when they talk about feelings.[8]

Naess resists a conception of reality as something endowed with an objective existence independent from the living being observing it, therefore he marginalizes the primary qualities that are supposedly intrinsic to the objects under observation. Rather, what a given living being experiences, what s/he sees or smells, touches or hears – the so-called 'secondary' qualities – are constitutive of reality itself. And this is even more true of the tertiary qualities – evaluative assessments like 'delicious' or 'encouraging', to use Naess' examples, that we may attribute to an object.[9] Thus, 'subjectivity' gains legitimacy as part of the world, and the constellation of relata involving a given being becomes essential to its ontology. In sum, according to Naess the qualities perceived by an observer in a given natural being are intrinsic to the being itself, and they form what he calls the world's 'concrete contents'.[10]

To discuss Ovid through Naess' position may seem a stretch on account of their distance in time, formation, role and intent: the danger of anachronism looms large. Still, one should stress that Ovid, too, looks at the living beings in his *Metamorphoses* as deeply embedded in the world they live in, their lifeworld (*Lebenswelt*). He also sees living beings in terms of a range of characteristics that correspond to the primary, secondary and tertiary qualities discussed by Naess. More specifically, the transformative events involving both the body of a human being and the body in which that human being will metamorphose stem from the dissolution of the so-called primary qualities.[11] This underscores the importance of the secondary and tertiary qualities in the constitution of reality, thus affirming the relational dimension of existence, whether human and nonhuman. Among the primary qualities, first and foremost it is shape that changes, thus fulfilling the poem's intent to sing 'of forms changed into new bodies'.[12] And often size changes too, from big to small and vice versa, as if the poet wanted to erase the value of the category of 'big' versus 'small' and reveal, so to speak, an ontology of the detail. For instance, the metamorphosis into men and women of the stones thrown by Deucalion and Pyrrha involves not only an increase in size but also an attention to the analogies between old and new body-parts (a constant of Ovid's narratives): the minute 'veins' of stones thus morph into those of the new-fangled generation of human beings.[13] Movement changes too, because many characters undergoing metamorphosis lose or transform their capacity for locomotion, either completely or partially.[14] When human beings are changed into trees, their roots tie them to the earth, their bark seals their bodies and faces. Io is transformed into a cow, unable to lift her head upward. Solidity too mutates, whether increasing, as in the case of some vegetal metamorphoses, or completely disappearing. Bodies often dissolve and become fluid, boundless beings such as rivers and springs.

In addition to the programmatic change of primary qualities, there is something striking about Ovid's accounts of physical transformations and about the resulting natural beings.[15] The 'subjective' qualities achieve significant importance. For instance, the poet's attention to colors is so dominant that Paul Barolsky claims, 'In his book we find a universe defined by color.'[16] But in the *Metamorphoses*, colors do not merely define the universe so that the poem gains an extraordinary pictorial quality. Rather, colors are often invested with emotional, personal meanings.[17] The laurel tree into which Daphne morphs retains the beauty Apollo saw in the young woman, now showing a warm glow.[18] Its newly acquired role in the religious landscape of the god expresses both the purity of the nymph and her denial of sexual love, while sublimating a distorted version of the god's desire. The plant's evergreenness, on the other hand, reflects (still according to Apollo) the god's own youth.[19] The wolf into which Lykaon is transformed continues to rejoice in blood, albeit of sheep rather than of human beings, and still retains the subjective, perception-related traces of the man: 'grayness the same, the same cruel visage, the same cold eyes and bestial appearance'.[20] In some cases, sound disappears once the human being is transformed into an inanimate being or a speechless animal. As soon as Niobe is transformed into a rock, her tongue becomes frozen to the roof of her mouth.[21] When Arachne becomes a spider, she stops speaking.[22] But in other cases a new voice conveys the emotions the metamorphosed being could feel before the physical change. The cry of the halcyon will forever reproduce the pain of Alcyone herself, when she saw her dead husband approaching the shore.[23]

In Ovid's imaginary world, metamorphosis occurs via alteration of those qualities that constitute the objective side of reality, but it also involves the qualities that constitute its subjective side – either maintaining or changing them. Metamorphosis always affirms an ontology of the natural world based on relations and perceptions.[24] Hence we should appreciate the relationality intrinsic in the perceptual qualities that inform Naess' ecological thinking and resonate in Ovid's poetics of metamorphosis. And we should also note the poem's environmental concerns along with the emerging invitation to the reader to embrace a moral approach to nature. Indeed, a subtle critique transpires from the theoretical framework of the poem, constituted by the description of the origin of the world in Book 1 and Pythagoras' speech in Book 15: the cosmos itself is a system of relations that is harmonious but vulnerable, while the earth is a natural being subjected to human (and divine) violation and has an integrity that requires protection.

This essay will pursue these environmental threads, starting from the beginning (i.e., the creation of the world) and moving to a discussion of relevant episodes of *Metamorphoses*. In the first section I will focus on Ovid's account of the world that human beings (and other creatures) live in, their lifeworld, looking at the problematization of human beings' approach to the earth and how it affects their approach to animals. Ultimately, the notion of kinship and particularly of motherhood will emerge as central to Ovid's discourse, which features the earth itself as a living being. Next, I will turn to the episodes where stones transform into human beings and Dryope becomes a lotus tree, focusing on the change and permanence of objective and subjective qualities and the ensuing relational ontology of the natural world.

The lifeworld of the *Metamorphoses*: from creation to exploitation and disclosure

An exuberant fusion of philosophical knowledge and mythical erudition, Ovid's account of the origin of the cosmos in *Metamorphoses* 1 follows a zigzag trajectory, in a chronological perspective that aligns him with the investigations into nature of the early Greek philosophers.[25] The world is created, disappears and emerges again, offering a living environment that will continue to be further diversified by a succession of metamorphic events that lead to the origination of plants and other natural beings, animate or not.[26] At first there is Chaos, a shapeless mass of discordant elements, which a god (or a better nature) organizes by dividing and allotting to their proper places according to their weight.[27] Then the earth is molded into a uniform globe and receives its various geographical features – streams and shores, swamps and lakes, large plains, valleys and mountain peaks – which will be mentioned in the course of the poem as specific settings for the various metamorphoses. For instance, Lykaon will turn into a wolf fleeing across the silent fields of Arcadia, while after the deluge Deucalion and Pyrrha will find refuge on the peak of Parnassus, where an entire race of humans will originate from stones. Syrinx, on the other hand, will become a reed, rooted on the bank of a river, when running away from Pan. Be that as it may, the catastrophic deluge ordered by Jupiter and aimed at destroying Lykaon and his impious contemporaries has the effect of bringing back the same indistinct situation of primordial Chaos. Now too 'there are no longer boundaries between earth and the sea, for everything is sea, and the sea is everywhere without a shore'.[28] Only when a pious couple survives – Deucalion and Pyrrha – and floods and streams are called back does the world start to emerge again, ordered as before but now utter desolate, bereft of human beings and any other life-forms.

In making a god responsible for the emergence of the world, Ovid seems to toy with the model of creation sustaining Plato's *Timaeus*[29] and with the idea, essential in that dialogue and still conspicuous in *Metamorphoses*, that the world is not complete unless it is inhabited by different forms of life.[30] So the account of the creation starts by describing a condition of lack, of the nonexistence of what has yet to become. The sun did not light up the earth, nor did the ocean embrace it; moreover, Ovid writes, 'although the land and sea and air were present, land was unstable, the sea unfit for swimming, and air lacked light'.[31] Here he uses a series of negative adjectives that evoke movements and activities that would have taken place in these areas had various life-forms existed. To receive life, the world first needs to achieve order and design, and god's most urgent operation consists in replacing strife with harmony.[32] Indeed, the separation of the elements mentioned earlier, by which aether leaps upward and air stands below it, above earth and water,[33] has the effect of dissolving the elemental fight intrinsic to Chaos, turning the world into a system of harmonious relations. For as aether, air, earth and water form actual and livable areas of the world, each element is in its place, and all are in harmony.[34] Note also that Ovid, still in line with the Presocratic philosophers,[35] divides heaven into five zones that correspond to five areas on the earth, and only two of these benefit from a temperate climate; the other three are *uninhabitable*.[36] In the end, biological diversity dawns on the world to complete its distinct yet related environments, and such a culmination suggests harmony within each habitat – that is, each life-form occupies its optimal living environment. Ovid tells us (*Met.* 1.98–103),

So that every region of the world
Should have its own distinctive forms of life
The constellations and the shapes of the gods
Occupied the lower parts of heaven;
The sea gave shelter to the shining fishes,
Earth received beasts, and flighty air the birds.

The human being is the last creature to appear, and whether created by god from his own divine substance or by Prometheus from a mixture of earth and water, it has a separate origin. This exclusive origin establishes the difference between humans and animals (the former, more intelligent, are created in the image of god to dominate the second!)[37] and foreshadows the problem of how humans should regard animals and the earth they live in, a concern that surfaces in the subsequent myth of the four ages of man and which is then central to Pythagoras' speech in *Metamorphoses* 15.[38]

Human behaviour toward animals and the earth generates a set of connected issues, made explicit in Pythagoras' speech.[39] Different generations – golden, silver, bronze and iron – come into existence, then human domination over animals gives way to their control over the earth and its exploitation.[40] Created to dominate all the other living beings, humans end up conquering the natural world and its creatures: strife effaces their primordial harmony, and violence destroys their pristine peace.[41] In the golden age, lack of laws and fear are emblematic of a humankind that is inherently virtuous and hence, in Ovid's environmental (and moral) interpretation of this myth, content to live in its space of existence. For he tells us that at that time human beings have not yet intruded onto mountaintops to cut pine trees and build ships in order to reach foreign lands: 'Men kept to their own shores.'[42] But no intrusion has yet been made into the earth either (*Met.* 1.140–2):

No rake had been familiar with the earth,
No plowshare had yet wronged her; untaxed, she gave of
Herself freely, providing all essentials.

A human presence is implicit in the mention of the 'plowshare' (*vomer*), but human life did not rely on the subjugation of animals, which at that time lived free from fear and danger in their own environments. Pythagoras will remind us of animals' freedom and tranquil life in Book 15, expanding precisely on the kind of peace that was characteristic of the golden age described in Book 1. For the wise man, peace informed also human beings' interactions both with animals and among themselves (*Met.* 15.132–9):[43]

That time long past, which we now refer to as 'golden,'
Was blessed in the fruit of its trees, and in its wild herbs,
And in the absence of blood smeared on men's faces.
In that time, the birds flew through the air without danger,
The fearless rabbit went wandering over the meadows,

And the fish was not brought over the hook by its credulous nature.
All lived without ambushes; none had a fear of deception,
And peace was everywhere.

What ultimately granted animals safety was a human diet based on wild berries and fruit, in addition to grain and wheat, which the earth eventually began to produce spontaneously without being tilled (and hence wronged).[44] How radically this mode of life differed from those that followed! In the iron age a cluster of moral evils took roots: greed triumphed and led this generation of humans to break their exclusive tie with the environment they lived in, to travel to foreign shores, and to take possession of the land by imposing boundaries on it (*Met.* 1.71–93):

Now ships spread sails, though sailors until now
Knew nothing of them; pines that formerly
Had stood upon the summit of their mountains,
Turned into keels, now prance among the waves;
And land – which formerly was held in common,
As sunlight is and as breezes are –
Is given boundaries by the surveyor.
Now men demand that the rich earth provide
More than the crops and sustenance it owes,
And piercing to the bowel of the earth,
The wealth long hidden in Stygian gloom
Is excavated and induces evil;
For iron which is harmful, and the more
Pernicious gold (now first produced) create
Grim warfare, which has need of both; now arms
are grasped in bloodstained hands.

The iron age is when humans begin violating the tops of mountains to cut down pine trees and turn them into ships (a novel human-induced form of chaotic distortion of the natural order). They also reach the interior of the earth, 'piercing to the bowels' to acquire iron and gold, hence inflicting to an extreme the type of violence on the body of the earth that had begun with tilling during the silver age (through the subjugation of animals).[45]

In Book 15, it falls to Pythagoras, the greatest authority on natural philosophy, to reveal the full extent of these actions and images. Pythagoras, whose 'inner sight exposed what Nature kept from human view',[46] calls the earth the best of all mothers, *optima matrum*.[47] While discussing the law of change, giving equal weight to macrocosm and microcosm, he also explains the mutation of the earth's conformation. He observes that new caverns open up and that old ones seal over, presenting the possibility that the earth itself might 'possess an animal-like nature and be alive' (*sive est animal tellus et vivit*).[48] Ovid plays with this idea and, perhaps under the influence of Empedocles, has Pythagoras allude to the possibility of the earth's breathing.[49] Earth possesses openings (*spiramenta*), through which she lets out her 'flames'.[50] Thus, if the reader's memory has endured up to this point of the narrative, when Pythagoras mention the fact that the earth might be a living being, all the language previously used in the poem to describe humans' violent intrusions and exploitation, from the top of the mountains to the surface of the ground and then deep into its bowels, acquires its full meaning. With all its geographical features, which Ovid took the pain marvellously to describe in *Metamorphoses* 1, as emerging under a god's design, Earth demands human respect. And one way to respect it is to return to life as it was in the ages of man in Book 1: humans should 'keep to their shores'.[51]

In *Metamorphoses*, there is an episode that, in line with Pythagoras' belief, presents the earth as a suffering being –and not because of human greed but because of a boy's stubborn ambition. When Phaethon obtains the Sun's consent to ride his chariot across the sky, he wreaks havoc on the earth. Ovid indulges in a detailed description of the effects of the youngster's catastrophic ride from the burning of earth's highest points and the cracking of deep fissures to the conflagration of trees, leaves, and crops, and even of entire mountains, to the dying of springs and the drying up of rivers.[52] In the end, Mother Earth is personified and, aware of her upcoming conflagration, apostrophizes Jupiter with a long, irony-triggering speech.[53] We read (*Met.* 2.272–81):

> Kind Mother Earth, surrounded
> By the sea and by the water of the deep
> And by her streams, contracting everywhere
> As they took shelter in her shady womb,
> Though heat-oppressed, still lifted up her head
> And placed a hand upon her fevered brow;
> And after a tremor that shook everything
> Had subsisted somewhat, she spoke out to Jove
> In a dry, cracked voice: 'If it should please you
> That I merit this, greatest of all gods,
> Why keep your lightnings back? If I must die
> Of fire, why not let me die of yours.'

Earth is a living being subjected to death. The mention of her 'womb',[54] which under Phaethon's ride is swollen, with all the streams flowing back into it, underscores her maternal nature, although now reduced to infertility and unable to generate again. The adjective 'kind' (*alma*) draws attention to Earth's nourishing role, which she continues to play *qua* original mother.

Mentioned by Pythagoras and evoked in the myth of Phaethon, the status of the earth as mother is central to the myth of Deucalion of Pyrrha and the metamorphosis of stones into a new generation of human beings. After the transformation of Lykaon, this is the second narrative in the poem to account for the metamorphosis of individuals.[55] It is unique with regard to the dominant direction of physical transformations, from the shapes of human beings into those of plants, animals and geographical features.[56] Here it is anonymous *stones* that metamorphose into human beings – so many stones as to populate the vast solitude of the world. On one level, this myth constitutes Ovid's witty reinterpretation of the conception of the earth as the origin of all life-forms, which was so dear to the Presocratics and resurfaced in Roman times with Lucretius.[57] Ovid follows this idea in the *Metamorphoses*, when he tells the emergence of nonhuman forms of life after the deluge.[58] But at another level, this narrative also makes it possible to weave into the poem the kind of change which Pythagoras so forcefully predicates as a universal phenomenon in Book 15. Up to this point (the transformation of humans into stones) change has involved only the movement of the elements – from Chaos into order, then back into the chaos caused by the deluge. Now we learn that all of those human beings who, in the course of the poem, will metamorphose into natural beings, living and not, are themselves already the products of a physical transformation.[59]

But even more importantly for this chapter, this myth turns around the meaning and the role of Earth, which is clearly central in Ovid's reinterpretation of the ages of man, then emerges in the account of Phaethon's tragic ride, and finally plays a role in Pythagoras' speech, with his reference to Earth as *optima matrum*.[60] In the episode of Deucalion and Pyrrha, which we will discuss in detail in the next section, Themis delivers an ambiguous oracle: they should throw 'the bones of the mother' behind their backs. This leads the pious couple (and the reader) to regard the earth as a mother,[61] – for the stones in the earth are called 'the bones of the mother'.[62] The couple shows reverence toward 'the mother's bones', handling them in a religious setting and with their heads covered. This seems to point to a model of respectful behaviour human beings should follow in their dealing with the body of the earth and its parts, or, in other words, with the natural world and its features.

Metamorphoses, motherhood and relational ontologies: From the earth to trees

Following the current of the Naess-inspired reflections at the beginning of this essay, let us now look at the interplay of qualities at stake in the physical transformations in the poem. We will understand what kind of ontology underpins these transformations. In the episode of the metamorphosing of stones into human beings, once Deucalion grasps the meaning of Themis' oracle, he proceeds, along with his wife, to obey the goddess' commands. Ovid writes (*Met.* 1.398–415):

> As they descend, they veil their heads and loosen up their robes,
> And cast the stones behind them as the goddess
> Bade them to – and as they did, these stones
> (you needn't take this part of it on faith,
> for it's supported by an old tradition) –
> these stones at once begin to lose their hardness
> and their rigidity; slowly they soften;

> and once softened, they begin to take on shapes,
> then presently, when they'd increased in size
> and grown more merciful in character,
> they bore a certain incomplete resemblance
> to the human form, much like those images
> created by a sculptor when he begins
> roughly modeling his marble figures.
> That part in them which was both moist and earthy
> Was used for the creation of their flesh,
> While what was solid and incapable of bending
> Turned to bone, what had been veins
> Continued on, still having the same name.
> By heaven's will, stones that men threw took the form of men,
> While those thrown from the woman's hand repaired
> The loss of women: the hardness of our race
> And great capacity for heavy labor
> Give evidence of our origins.

One may wonder why the stones do not metamorphose on their own but need the human touch to do so. On a narrative plane, it is again a matter of Ovid's inventiveness and love for surprising variations: the relatives of Prometheus, who is credited in the tradition with having created human beings (whether by moulding them out of clay or by means of exclusive gifts),[63] repeat the Titan's deed, but in an unprecedented way.[64] But human involvement in this creation is also in line with the relational frame that informs other episodes of metamorphosis in the poem. The stones thrown by Deucalion will become males, while those thrown by Pyrrha will become females. Relation triggers change, and the specific contact – whether with a man or a woman – determines gender. As for the metamorphosis itself, it unfolds through a change of the so-called primary qualities – solidity, shape and size – in addition to, arguably, movement: these mineral objects become softer and larger and take on forms so that, at this point, they resemble the unfinished works of a sculptor. Through this remark, Ovid continues the implicit parallel of this myth with Prometheus' creation of man. Soon, the finished products of the metamorphoses will acquire movement: the men and women deriving from the stones will actually live and move.[65]

Still, despite these changes, an implicit connection ties the bodies of the stones to those of the new human beings. As in other metamorphic events, Ovid reveals a hidden continuity, by bringing to the fore a set of bodily correspondences that involve secondary and tertiary qualities. The moist and earthy parts of the stones become flesh and solid bones, he tells us, while the veins remain the same. In the episode of Pygmalion's statue becoming alive, human veins must merely start pulsating.[66] But in the myth of Deucalion and Pyrrha the overall hardness of the stones subsists in the character of the new human kind, which is essentially hard-working and enduring.[67] Thus, a secondary tactile quality that depends on the perceiver – hardness – morphs into a quality of character that reflects human beings' nature. Granted, hardness seems to predicate an ambiguous trait that contributes to a given being's intrinsic (bodily) solidity, while also relating to the perception and evaluation by another party. Still, hardness explains the essence of this new generation of humans in subjective terms.

In the end, one may wonder about the ideological weight of the earth/mother conceit and how much emphasis Ovid intended to give to it. True, the Roman poet is no natural or environmental philosopher, and the syncretic philosophy interwoven in Books 1 and 15 is working toward the poetic creation of a marvelous and fluid world.[68] But in a poem that speaks of a continuous change investing macrocosm and microcosm alike and even affecting, so to speak, the flow of the narrative,[69] the author's concerns with the earth – her role as mother and her status as both living being and a living environment, sometimes violated by human beings (and by the gods)[70] – are there to stay. However playfully narrated, the constancy of these issues invites the reader to reconsider her relation with the earth in terms of kinship and in tune with the kind of connections Pythagoras stresses with regard to animals. The earth itself may be a living, breathing being.[71] When read in the context of the poem and not in isolation, then, the account of a human generation rising from stones seems to encode a subtle message.[72] In short: if Earth is their mother, human beings should live in harmony with her, and not desecrate that most intimate bond. They should not act like the evil people of the iron race who desecrate the earth and, at the same time, break family bonds. A son consults a soothsayer to know 'when he will change from heir to owner'.[73] Since the very existence of human beings stems from parts of the earth's own body, the way to achieve this harmony is to continue to feel, and be part of the original environment that gave them 'birth', and that they still inhabit.

Motherhood is also at the centre of another episode of metamorphosis in Book 9, that of Dryope into a lotus tree.[74] This time, motherhood compels us to question the ontology of the vegetal world, to problematize the approach one should hold toward plants, which complements Pythagoras' speech.[75] The episode is a fusion of poetic virtuosity and the grotesque: an idyllic landscape suddenly becomes a horror scene. It explains Dryope's metamorphosis through a flashback to the previous metamorphosis of another nymph and weaves together, along with similar yet contrasting destinies, a variety of themes – rape and escape, ignorance, innocence, and injustice. Dryope possessed outstanding beauty, was raped by Apollo, found a human husband, and gave birth to a child. During a stroll with her infant, she reaches 'a lake whose banks are crowded with myrtle, whose shores slope gently down to meet the water' and picks some flowers from a lotus plant on the shore to entertain her child: a doomed move! Drops of blood fall from the flowers as the branches of the plant quiver with fear. Suddenly, Dryope starts metamorphosing into a plant from her feet upwards, while still feeding her child on her breast. Her form changes, and likely so does solidity, which increases, and movement, which vanishes. Her sister's embraces succeed in delaying her metamorphosis, only to make the tragic process even more excruciating. Dryope did not know that the lotus plant she tore flowers from was formerly Lotis, a nymph, who, sharing the same destiny as Daphne and Syrinx, escaped from the lust of her pursuer by becoming that plant – this we learn from her sister. While undergoing the transformation, Dryope addresses father and husband, newly arrived at the scene, She grieves for her innocence and admonishes them for the future:

> I have done nothing to deserve this evil!
> I bear a punishment that has no crime!
> I've lived a blameless life –and if I lie,
> Then let my foliage shrivel and die
> And may the axe and flame consume me quite!
> But take my baby from his mother's limbs
> And give him to a wet nurse who will bring
> Him here and nurse him underneath this tree,
> And let him play here too; when my poor babe
> Has learned to speak, let him come to this place
> And greet his mother here and sadly say,
> 'My mother is hidden underneath this bark.'
> Let him be frightened of the lake, and mindful
> Never to pluck the flowers from the trees,
> And think that in each bush a goddess hide!
> Farewell, dear husband – sister – father!
> If you can show compassionate respect,
> Than keep the sharpened pruning hook far hence,
> And guard my foliage from browsing flocks.[76]

Dryope's own motherly gesture, namely the simple act of picking flowers to entertain her child, proves to be fatal because, without knowing it, she violates the life of the nymph hidden in the lotus. This turns her into a plant herself: the new vegetal body into which she is transformed is determined by the flowery bush she has violated. In describing the process of metamorphosis, Ovid bring into focus the parallels between human bodies and plants, which he has meticulously detailed in the first example of botanical metamorphosis of the poem, that of Daphne into a laurel tree. The skin becomes the bark, the arms branches, the hair foliage, and the feet roots.[77] But in Dryope's case, the emphasis is on the young woman's dramatic resistance to the metamorphosis, and on her absolute powerlessness. She is becoming a tree, but before the bark seals her mouth forever, she reveals the wisdom she has acquired through the experience of her own transformation. Every plant may contain the secret life of a nymph so her child should be instructed to never pluck flowers and not to violate the bodies of trees, because they could be home to a goddess. Although this statement may seem to pertain only to the trees rising along the lake where the metamorphosis takes place, it is so general as to transcend it. All trees may conceal a secret life!

Now, the readers of *Metamorphoses* know that trees may contain the life not only of mortal women but also of human beings of all types, young and old, females and males.[78] According to Dryope's instructions, her child should be brought to the lotus tree she has now become, should recognize his mother, and should therefore approach that vegetal being with the same filial respect and affection he would have had for his human mother, had she not been transformed.

A similar admonition is given to the husband and the father: they should treat the trees of the lake with 'compassionate respect' (*pietas*) and keep the flocks away. *Pietas* indicates the respect human beings should have for other human beings, particularly their parents.[79] This sentiment is projected onto the relation of Dryope's husband and father to the tree she has become (and all other trees as well). This highly significant attitude, *pietas,* also appears elsewhere in the poem. Its disappearance contributes to the ultimate moral collapse of the iron age, when it is said to 'lie vanquished'.[80] *Pietas* is also involved in the mutual devotion of Deucalion and Pyrrha, in their approach to the gods and their handling of Mother Earth's bones.[81] Whereas Dryope's form changes (and so do her movement and possibly her size), the relations and relational modes of her previous existence must endure in the new vegetal body. She must be treated as a mother.

In the end, we are compelled to ask a question as to what types of beings are the plants, when they result from the metamorphosis of human beings, and what capacities they possess. I have discussed this question elsewhere.[82] To conclude this essay, I would like to stress a point which is especially relevant to the ecological reading of Ovid's *Metamorphoses*. Once the hidden nature of trees is revealed, one should feel a filial relation with them and behave accordingly. So it is, we have seen, in the story of Deucalion and Pyrrha in respect to the earth. Beyond Ovid's playfulness in crafting his narratives, the poem conveys a relational ontology and warns against the objectification of the natural world so dear to the physicist. The stories I have discussed encourage us instead to create subjective and personal ties between us, as observers, and the environment in which we live, namely the earth and its trees. And the revelation regarding this prescriptive behavior surreptitiously complements Pythagoras' discourse on our kinship with animals.

Notes

1. The expression 'deep ecology' appears for the first time in Naess 1973. It defined Naess' philosophical ecology as one grounded in principles of 'diversity, complexity, autonomy, decentralization, symbiosis, egalitarianism, and classlessness' in contrast to a shallow ecological vision that fights against 'pollution and resource depletion' in view of the 'health and affluence of people in developed countries' (95).
2. Schilpp 1949: 5.
3. Naess 2008: 61, which continues: 'There are worlds of minerals, rocks, rivers and tiny rivulets, plants, hardly visible animals or big ones (larger than a centimeter), plant and animal societies, tiny and great ecosystems – all more or less easily available for enjoyment, study, and contemplation. The meaningfulness inherent in even the tiniest living beings makes the amateur naturalist quiver with emotion. There is communication: The "things" express, talk, proclaim – without words"' (2008: 61).
4. Naess 2008: 61.

5. Communication and language interpretation was indeed a focus of Naess' philosophy, which he addressed in Naess 1953 and 1966.
6. I define human beings' transformative outcomes as 'natural' because, although they are often due to a god's intervention, they tend to unravel analogies between the bodies of humans and animals' or plants' hence presenting metamorphosis as a phenomenon that possesses a 'physical rationale'.
7. Naess 2008 (originally published in 1985 in *Inquiry* 28: 417–28). The essay sets out to defend those environmental activists who in their fight to defend natural beings such as a river, a wood, or a kind of animal are often criticized for expressing mainly 'feelings and subjective likes and dislikes' (2008: 70).
8. Naess 2008: 70–1.
9. The distinction between primary and secondary qualities goes back to Democritus (see DK 68 B 9), 'by convention sweet, by convention bitter, by convention hot, by convention cold, by convention color, but in reality atoms and void,' trans. by Taylor 1999) and was subsequently elaborated by Galileo and then Locke. Primary qualities (i.e., shape, size, motion, solidity) are considered to be qualities intrinsic to bodies, while secondary qualities (i.e., colours, odours, tastes, felt textures) are causal powers that result from the primary qualities and produce various sensations in us (for Galileo, see Martinez 1974; for Locke, see Uzgalis in the *Stanford Encyclopedia of Philosophy*). As for tertiary qualities, they are qualities of a thing as objects of evaluation.
10. Naess 2008: 72–3.
11. In Ovid's poem, physical transformation usually affects human beings who become other natural beings. An exception is the reverse metamorphosis of the stones thrown by Deucalion and Pyrrha that become human beings, a myth that represents a variation of the theme that life was born from the earth (discussed further below).
12. Ov. *Met.* 1.1–2.
13. Ov. *Met.* 1.569–70.
14. See Solodow 1988.
15. If, as announced in the proem (*Met.* 1.1–2), the change of form(s) is central to the poem, it attracts the changes of other so-called primary qualities. For instance, the human being that becomes a tree also changes with respect to movement and solidity; see above.
16. Barolsky 2003a: 55.
17. For the colours of white and red in erotic contexts, and particularly the myth of Pyramus and Thisbe, see Rhorer 1980.
18. Ov. *Met.* 1.762.
19. Ov. *Met.* 1.563. Admittedly this comment by Apollo continues in a humorous way the self-referentiality the god has manifested throughout his approach to Daphne while it reproposes the analogy between human hair and foliage activated in many episodes in which humans beings and nymphs metamorphose into plants.
20. Ov. *Met.* 1.327, 329–32. Interestingly, with a playful twist, while the colour of Lykaon remains, that of the wolf Daedalion changes, once he is transformed into a stone (Ov. *Met.* 11. 405–6).
21. Ov. *Met.* 6.303–12.
22. Ov. *Met.* 6.140–5.
23. Ov. *Met.* 11.716–35.

24. I connect these two notions because I consider perceptions a first instantiation, and basis, of relations.
25. As Aristotle reports, the early Greek philosophers accounted for the existence of the universe (*to holon*) down to animals and plants indicating what moves it (whether strife, love, mind, or spontaneity) and what the nature of matter is (see Aristot. *PA* 1 640b5–13). The reference to the 'motive origin' reveals that he is alluding to the theories of Empedocles, Anaxagoras and Democritus.
26. For the cosmological implications and etiological nature of many metamorphoses, see Myers 1994: 27–70.
27. The process of formation of the ordered world by separation of the conflicting elements has its origin in Anaximander (see DK 12 A 9); see Guthrie 1962: 78–9.
28. Ov. *Met.* 1.404–7.
29. Fantham 2004 mentions this connection, without discussing in detail the key points of contact between *Metamorphoses* and *Timaeus*.
30. For the god as creator, see Plat. *Tim.* 28C, 29A; and for the world as the living place of creatures, see, for instance, 33B: 'And he bestowed on it the shape which was befitting and akin. Now for that living creature which is designed to embrace within itself all living creatures the fitting shape will be that which comprises within itself all the shapes there are'; and 41B: 'Three mortal kinds still remain ungenerated but if these come not into being the Heaven will be imperfect; for it will not contain within itself the whole sum of the kinds of living creatures, yet contain them it must if it is to be fully perfect.'
31. Ov. *Met.* 1.18–20. All translations from Ovid's *Metamorphoses* are those of Martin 2009.
32. For an allusion in this passage to Love and Strife as the two forces that unite and divide the elements in Empedocles' doctrine, see Della Corte 1985: 9.
33. Ov. *Met.* 1.33–9.
34. Ov. *Met.* 1.32. Note how this outcome reflects Naess' definition of *ecosophy* as 'a philosophy of ecological harmony and equilibrium' (1973: 5).
35. The first to be credited with a distinction of the world into five zones is Parmenides of Elea, see DK 28 A 44a; with Rossetti 2017.
36. Ov. *Met.* 1.60–70 (*non est habitabilis*). Interestingly, the division of zones will return again in the myth of Phaethon and his tragic journey riding the chariot of the Sun (Ov. *Met.* 2.129–32).
37. See lines 76–7 (*sactius his animal mentisque capacious altae deerat adhuc et quod dominari in cetera posset*). The separate origin of human beings, on the one hand, and animals, on the other, characterizes also the emergence of life after the deluge (see below), showing this to be a constant in Ovid's conception of the origin of life. With this dual model he departs from the Presocratic philosophers as well as from Lucretius, and his view of the human being as god's creation is close to Plato's *Timaeus*. For the Stoic influences on the view of man as a *sanctius animal* in the image of god, see Alfonsi 1958. It bears noting that this division is playfully reflected in Morpheus' exclusive capacities to impersonate human beings in sleep while another god simulates the animals and a third one 'soil, rocks, waves, and tree trunks' (Ov. *Met.* 11.633–42).
38. Scholars have considered the role of Pythagoras' speech in relation to the metamorphic episodes of the poem noting its closeness, for the philosophical mode, to the cosmogonic account of the beginning (see, for instance, Segal 1969b: 262; Della Corte 1985: 4; Myers 1994: 27). In the following pages I too consider Pythagoras' speech as a pendant to the initial cosmogonic account, but unlike previous studies, I show how they complement each other with regard to the ecological threads of the poem pursued by this essay, i.e., the world encompassing different habitats as *Lebenswelt* with the consideration of the earth as a living

being and mother, subjected to human violation along with her nonhuman creatures in their space of existence.

39. In his condemnation of humans' meat-eating, Pythagoras ultimately attributes humans' mistreating of animals to their neglect of the earth and its fruits, that is, to their disregard for a vegetarian diet (*Met.* 15.91–126); the association of diet and peace is made also in the discussion of the myth of ages (see Ov. *Met.* 1.123–4).
40. Admittedly, humans' domination of animals is mentioned not in the myth of the ages of man, but in the first creation of man that precedes it. Still, we can consider it implied in the rise of agriculture during the silver age.
41. Ov. *Met.* 1.123–4. In Ovid's version of the myth of races, human beings' intrusion and control of the natural world may allude to Roman ideology of expansion and conquest, hence pointing to a symbolic association between control of nature and subjection of territories and people that surfaces over and over in imperial writers (see, for instance, Campbell 2012, chapter 10).
42. Ov. *Met.* 1.135.
43. The last two lines of this excerpt echo Empedocles' B 130, cf. Fabre-Serris 2018: 312.
44. Ov. *Met.* 1.104–6 and 109–12.
45. Ov. *Met.* 1.123–4.
46. Ov. *Met.* 15.63–4.
47. Ov. *Met.* 15.91.
48. See *Met.* 15.342 (*sive est animal tellus et vivit*). The question of whether the earth is a living being, provided with sensation or merely an inert body, also appears in Seneca's *Natural Questions* (2, 4).
49. Empedocles spoke of the sweat of the earth (DK 31 B 55, see Zatta 2018) and claimed that all things breathe (DK 31 B 100).
50. Ov. *Met.* 15.345; see Hardie 2015: 527.
51. Ov. *Met.* 1.135.
52. Ov. *Met.* 2.210–59.
53. Interestingly, in Latin poetry the earth may feature as a living being who is apostrophized. In Seneca's *Thyestes*, for instance, after knowing from Atreus about the macabre banquet Thyestes addresses the earth asking her why she does not open and swallow him and his brother and the entire Mycene into the realm of death but remains instead motionless, 'just a solid mass' (*ignavum pondus*) (Sen. *Thiest.* 1006–21). The difference with Ovid is striking because in the Senecan tragedy the earth is called to be a witness of the abominable cannibalistic feast and to act against it triggering a catastrophic event that is proportional to the perpetrated crime while in *Metamorphoses* the earth *is* the witness of the violence she is undergoing and, aware of her imminent death, asks Jupiter to intervene.
54. Cf. the origin of nonhuman forms of life after the deluge, which Ovid describes in terms of the earth's begetting (*pario*) (Ov. *Met.* 1.417).
55. In fact, the transformation of Chaos into an ordered world too is made to fit the description of metamorphosis given at the outset of the poem as 'forms changed into new bodies'. Chaos has a 'vultus' (Ov. *Met.* 1.6) that becomes diversified by separation into the bodies of the sky, the earth and the ocean, but the myth of Deucalion and Pyrrha inaugurates the type of linear metamorphosis prevalent in the poem from an embodied form into a new body.
56. For a discussion of the connection between metamorphosis and natural science, see Myers 1994: 28–30.

57. For the Presocratics, see DK 12 A11, DK 12 A30 (Anaximander); DK 21 B 33, DK 21 B 9 (Xenophanes); DK 60 A 1 (Archelaus); DK 31 B 62 (Empedocles); DK 68 B 5 (Democritus); for Lucretius, book 5 of *Rerum Natura* with Campbell 2003.

58. Ov. *Met.* 1.416–21.

59. Even if they are offspring of human parents, the human beings that metamorphose in the poem ultimately descend from individuals who have emerged from stones. The case is different for nymphs. They are demi-divine creatures whom, in the scheme of the poem, Jupiter tries to protect from the cruelty of the preceding generation of human beings born from the blood of the giants, contemporary of Lykaon, and destroyed by the deluge (Ov. *Met.* 1.192–5).

60. Ov. *Met.* 1.91.

61. Much has been written about the revelational nature of Pythagoras' speech, its Platonic resonances, and the Empedoclean model through Lucretius (1.731–3, for these and other intertextual references, in addition to relevant bibliography, see Setaioli 1998: 500–4). Yet attention should also be paid to the revelation *in fieri* at stake in the episode of Deucalion and Pyrrha with the oracle of Themis and its interpretation, the core of which turns around the earth as mother and thus ties in with Pythagoras' speech, underscoring a crucial environmental thematic concern in *Metamorphoses* and connecting Book 1 with Book 15.

62. Ov. *Met.* 1.393–4 (*magna parens terra est, lapides in corpore terrae ossa reor dici*).

63. See respectively Apollod. 1. 7. 1; Hes. *Theog.* 565–6, Aeschyl. *Prom.* 447–506.

64. The myth of Deucalion and Pyrrha with the emergence of human beings from stones is in fact of ancient origins. Mentioned in a fragment by Hesiod (Merkelbach and West 1967: 234), it appears in Pindar in association with the flood (Pind. *Ol.* 9.42–54; see Griffin 1992: 39–40), but both versions are tied to the Locrian people in central Greece. By contrast, Ovid universalizes this account, severing its ethnic relevance and using it to explain the creation of a new generation of human beings after the quasi-extinction caused by the flood, and thus presenting all human beings as 'children of the earth'.

65. The transformation of Pygmalion's statue into a woman provides an interesting supplement to what Ovid leaves unsaid in this myth, albeit in a different context (*Met.* 10.243–97): he focuses not on the actual awakening to life as in the myth of Pygmalion, but on the analogies connecting the bodies before and after the metamorphosis.

66. Ov. *Met.* 10.283. In fact, arguably the veins remain the same with respect to the new shapes the stones have achieved. Up to that point, they too must have changed shape in order to become a part of the human body.

67. Ov. *Met.* 1.414–15.

68. Books 1 and 15 interweave numerous resonances of Presocratics beliefs. In Book 1, to Empedocles' theory of the elements (see, for instance, DK 31 B 6) one should add the process of organization of Chaos into the ordered world by separation, which seems to echo Anaxagoras (DK 59 B 13). In Book 15, on the other hand, Pythagoras fuses in a continuous narrative the theory of metempsychosis, attributed to him (DK 21 B 7), and the Empedoclean revelation that death does not exist (DK 31 B 8) with Heraclitus' belief that everything is in flux (DK 22 B 12).

69. In this respect, see Barchiesi 2005.

70. With the approval of the other gods, Jupiter destroyed the earth as a living environment in order to annihilate the impious human race inhabiting it.

71. See above.

72. This interpretation raises the question, outside the scope of this essay but inherent in the reading it proposes, of how one should read *Metamorphoses* – that is, in a linear way or through cross-references. The poem seems so constructed as to elicit the second and to change the reader's reading process itself.
73. Ov. *Met.* 1.148.
74. In fact, motherhood features as a motif of this book, which early on accounts for Alcmena's inability to give birth on account of Juno's hostility.
75. In supporting vegetarianism Pythagoras does not discuss humans' relations to plants (see Sissa 2019), which the myth of Dryope does address.
76. Ov. *Met.* 9.373–93.
77. Ov. *Met.* 1.547–51.
78. See, for instance, the episode of Philemon and Baucis in *Met.* 8.611–724 and that of Myrrha in *Met.* 10.298–519.
79. For a synthetic discussion of the meaning of *pietas* from Lucretius to Aquinas, see Garrison 1992: 9–14.
80. The epigrammatic sentence, *victa iacet pietas*, concludes the excursus on the moral decline of the iron age; see Ov. *Met.* 1.149.
81. See *Met.* 1.371–99.
82. See Zatta 2016, 2020.

AGRICULTURE

CHAPTER 9
LANGUAGE, LIFE AND METAMORPHOSIS IN OVID'S ROMAN BACKSTORY

Diana Spencer

Introduction

Ovid's *Metamorphoses* is a landmark poem that challenges norms of genre, reconceptualizes myth, and gives new Latin form to the human body within an environment where shapeshifting is commonplace. This is a poem in which there are some 250 acts of metamorphosis in 12,000 lines of text.[1] In this chapter I investigate the affective emphases and environmental entanglements centred on transformation that emerge from a journey into its prehistory: this is an exploration of the literary imaginary that fashioned Ovid's epic and in which context its allusive texture should be understood. From a twenty-first-century Anglophone and Western perspective, Ovid's dangerous, delightful, frightening and marvellous environmentalism is both strange and curiously inviting, but rather than emphasizing its uniqueness, this chapter asks to what extent is it exceptional and explores how rooted it is in the sensuous perspectives of the world from which it emerged. This chapter therefore gives the *Metamorphoses* a genealogy, tracing its precursors from an intellectual model emerging as Rome's territorial empire expanded and in doing so, put increased scrutiny on the city's roots on the Seven Hills by the Tiber.

Evidence from Cicero – our most prolific extant author from this era – makes abundantly clear that Roman bodies understood in the widest sense (individuals, families, the State; alongside the fabric of the city itself and its cultivated territory) were conceived as subject to metaphorical and physical transformation for much of the century before the Augustan Principate. As this chapter discusses, he was not alone in this perspective. Authors before Ovid produced nothing in Latin explicitly centred on metamorphosis, on the scale or ambition of Ovid's epic, but the intellectual and ideological possibilities opened up by the concept were already vivid in works of the generation formed by the turbulent early decades of the first century BCE. Late Republican interest in transformation repeatedly intersects with questions concerning teleology and the place of humans, especially Romans, within an endlessly mutable environment, and finds echoes in what contemporary thinkers have dubbed *natureculture* in the wake of Haraway's coinage.[2]

Politicians, thinkers, and authors were still consciously indebted to the idea of Rome as uniquely in tune with its topography and environmental situation. Simultaneously, more experience of diverse entities and new landscapes redefining Rome's 'limits' was encouraging intellectuals to reconsider the foundations of Roman environmental awareness in the light of new tensions caused, and spaces opened up, by territorial ambitions and encounters with various kinds of Other.[3] In the context of Rome's empire building, Parthenius'

Metamorphoses shows how this theme was bubbling up through intercultural products of conquest and slavery gaining expression at Rome.[4]

To gain an understanding of these complex patterns in the Roman backstory out of which Ovid's epic emerged, this chapter works through three case studies. First, an associative reading linking sections from Virgil's poem of Roman farming life (*Georgics*), Cicero's second speech on agrarian reform (*De Lege Agraria*), and Lucretius' natural-scientific epic on the nature of everything (*De Rerum Natura*); next, an exploration of the metaphoric and practical resonances of grafting versus propagation from seed, in the agricultural study written by the era's great polymath, M. Terentius Varro. My final case study takes us back to Virgil in a comparative reading of *Eclogue* 6 with Catullus' miniature epic poem 64. Both works have an experimental quality, and in their complex approaches to ekphrasis and world building, using mythic and prophetic figures amidst richly connected landscapes, we most closely approach the cosmogonic qualities of Ovid's overall scheme.[5] In the nuances of this third case study we see most vividly the places where the mysteriously tangible and the inexplicable meet, a type of convergence strikingly outlined by David Abram's *The Spell of the Sensuous*.[6]

My selection of ancient case studies shows how animate, sentient and active participants in the cosmos shed light on different kinds of discontinuity and asymmetry in human/non- human relations. They also manifest some of the strange and wonderful in those symbioses which are invisible to or inexpressible within the domain of science. These works, and others, open up a number of connected issues against which Ovid's epic of shifting scenes and mutating materiality takes shape.[7] Environmental imagination, at Rome in this era, was still vigorously rooted in the idea of natural, seasonal cycles. Yet by the mid Republic, economic experience had diverged from other urban centres in antiquity, transitioning from (quoting Seth Bernard) 'a Weberian farmer-citizen city to [...] a more urban-centred working population'.[8] This shift led to a flourishing intellectual trend in investigating the developing urbanism of Roman citizen identity, its impact on what 'now' might constitute an appropriate sense of self in a Roman universe, and how to systematise transforming identities within a highly traditional model of citizens as a corporation (*res publica*) living in line with precepts derived from a legendarily agricultural origin story (*mos maiorum*).[9] Alongside heuristic models employed in Stoic and Epicurean philosophy, ancient Roman citizens also continued to dwell in a world structured and given meaning by civic religion.

Civic religion, with its rituals, cult practices, and local or special-group spinoffs, acknowledged the perceived contingency of divine relationships and forces invisibly animating entities such as trees, waters, mountains, fire, air, and earth in the material world.[10]

Case study 1: Biomes (Virgil *Georgics*, Cicero *De Lege Agraria* and Lucretius)

Agricultural identity was deeply embedded in the patriotic mythmaking and storytelling around which productions of Roman civic identity clustered. This first case study looks at how the life of the farmer and a citizen's relationship with the land were endemically concerned with cycles of transformation and questions of teleological instability. Drawing on examples of environmental thinking from Virgil, Cicero and Lucretius, we move from farming and territorialism through to cosmogony and the nature of perceptual reality, tracking how these themes were emerging as cultural touchstones against which *Metamorphoses* should be read.

Virgil's poem of the farmer's life, written in the 30s BCE (a generation or so before Ovid's epic),[11] warns that Rome's agricultural heart beats within a complex array of agencies, all operating with distinctively embodied and brokered relationships. Radical transformations take place within this environment, but need to be understood as functions of wider and sometimes unintelligible patterns. Read in this way, the *Georgics* models an imaginary that is significantly implicated in underpinning the *Metamorphoses* as a document of lived experience and shared culture. This Virgilian model is especially important when considering the careful attention Ovid pays to the symbolic and metaphorical links between the sites of transformation and their relationship with the transformed characters. We can see something of this programmatic aspect in these early lines from Book 1 of Virgil's poem:

> Care should also be taken to learn in advance
> the customs [*morem*] of places, and the cultivation and habitude [*cultusque habitusque*] appropriate to their lineage,
> and what each region bears and what each refuses.
> [. . .]
> Uninterrupted, these laws and eternal pacts [*leges aeternaque foedera*], in certain lands, Nature has established [*imposuit natura*]. Right from the time when first Deucalion cast stones out into the empty world,
> Whence men were born, a hard race.
>
> *Georg.* 1.51–3, 60–3[12]

Virgil's choice of language ramps up the anthropomorphic and adversarial qualities of environmental imagination but also makes space for different ways of conceiving what it means to be human: respectful collaboration and agreed models of cultivation, but also a recognition that humanity is materially and filially a function of powerful Nature's oversight of the cosmos.[13] His careful emphasis on humanity's inherent intermateriality, with first forebears emerging from a punitive environmental catastrophe into a regularised system of symbiosis and acculturation, shows how metamorphosis has deep roots. He invites Roman citizens to imagine themselves as uniquely qualified by nature and culture to understand and transform, as well as be shaped by, their own territory.[14] The practical implications of this imaginative worldview were already part of the escalating political volatility, as is evident early in Cicero's three speeches *De Lege Agraria* delivered in 63 BCE at the beginning of his consulship, and attempting to defeat agrarian-reform legislation proposed by P. Servilius Rullus, Tribune of the People. In his second speech, a disorienting *tour-de- force*, he sets out to inspire the popular assembly against a transformative piece of legislation that might well have improved their lives. Cicero asks:

> Quid enim est tam populare quam pax? qua non modo ei quibus natura sensum dedit sed etiam tecta atque agri mihi laetari uidentur. Quid tam populare quam libertas? quam non solum ab hominibus uerum etiam a bestiis expeti atque omnibus rebus anteponi uidetis.
>
> *For what is there so fitting to the people as peace? in which not only those to whom nature has given sense, but even the houses and fields seem to me to take pleasure. What is so fitting to the people as freedom? which not only by mankind but even by the beasts is sought out and prioritised above all else, as you see.*
>
> *Leg. ag.* 2.9

The reform seems at least in part to have responded to Rome's imperializing acquisition of new territory, how that territory was (or was not) removed from the use of its original occupiers, and increasing demand for land and growth in agribusiness as peninsular Italy stabilised politically. Its aim was a new distribution of the traditional *ager publicus*, taking from the large-scale farmers and aristocrats who had in some areas (as well as in the popular imagination) gained a monopoly over vast tracts, and a resettlement of the urban and rural poor and landless to form new agricultural communities and to found new towns.[15] Cicero successfully transformed the bill into an attack on civil liberty and the values that had made Rome great.

In the quoted passage, any sort of transformative rebalancing of how the land is owned and by whom it is cultivated is anathema to the peace which victorious Romans have fought for, and will destroy the traditional values of Roman identity and the culturally conceived norms of what a citizen most authentically is.[16] This is not just about the 'animate' community, '*quibus natura sensum dedit*', indebted to nature for perceptual and sensory capacity, but also the Roman-engineered environment of the *longue durée*: buildings and agricultural fields. Although Cicero is not articulating it explicitly, his language ('*tam . . . quam . . . non modo . . . sed etiam . . . tam . . . quam . . . non solum*') taps into the idea that while the divide between Romans, their works, and the natural environment is full of blurred lines and porous zones of contact, these active and complex intersections through which *natureculture* is understood to produce varieties of Roman identity need to be maintained in their different and traditional forms of integrity. Cicero's perspective in this speech perhaps unconsciously stakes out a position which acknowledges the risks inherent in an imperial model of relentless environmental and socio-ecological devastation and upheaval. As Kyle Harper has observed, 'the Romans were, in planetary perspective, lucky' because their empire building coincided with (and was significantly enabled and aided by) environmental conditions that also made for the stable trade conditions and agricultural growth.[17] It was this model, as Harper argues, that was simultaneously a marker of Rome's monumental success in expanding and a contextual condition for eventual decline.[18]

Behind Cicero's argument and the idea of Rome as an empire there also sits a tradition we have already encountered in Virgil, *Georgics* 1: understanding peoples and places, climates and characteristics, as intelligible only as part of an ethically calibrated environmental ecology of existence, within which the right people are understood to be matched to the appropriate territory. Here, Cicero's sweeping demand of Rullus: '*cur eos non definis neque nominas*' (why neither define nor name the lands, *Leg. ag.* 2.66) develops into a determined emphasis on the underpinning existence of essential characteristics for each tract of land in question, for the absence of detail on which, in this context, he berates Rullus: each *locus* or 'site' requires a definition and also a summary of its inherent qualities (*natura*) as field land (*ager*).[19] These traits will shape the identity of the success of the land and its peoples, the farmer and the community, and the kinds of 'companions' (animal, vegetal, human, technological) with whom he will collaborate.[20] Yet for Cicero's purposes, Rullus seems not to understand this basic premise whereby a rightness in alignment of peoples and place is crucial, nor to care what the nature or location of the land up for distribution might be, nor how its assimilation to a new kind of Roman ownership will transform people and *res publica*.

As a social-engineering project, Cicero emphasises, this puts the proposed beneficiaries, the contemporary urban poor and the dispossessed, in a newly vulnerable position. To paraphrase Cicero, trying to transform them into the successful farmers of nostalgic history risks their starvation if transplanted as anomalies into the barren, unproductive lands eagerly off-loaded by the wealthy.[21] Cicero's argument thus frames Rullus' bill as what we might now term a challenge to the *res publica* understood holistically as a biome. A biome is a space defined by the features of its community, shaped by similar constituent animals, plants and other life forms, in tune with a prevailing climate and patterns of behaviour and habitation. When understood correctly by its inhabitants it makes possible a kind of 'common-sense' environmental awareness.[22] This reading sees the core identity of the *res publica* weakened and transformed when some constituent members are dispersed across lands whose compatibility and capability to collaborate with them is untested and indeterminate. Cicero's arguments often succeed precisely because they take contradictory positions. Hence, some members of the audience might even see the legislation as a direct attack on the urban commons who are to be bamboozled into support, before their redefinition by Rullus as human waste (*sentina*) to be drained off (*exhaurio*), as if Rome were again a primordial swamp.[23] This threat gains momentum as Cicero emphasizes the 'common-sense' argument that deceitful Carthaginians, hardy Ligurians, and luxurious, arrogant Campanians, all became so because of the nature of their biomes ('*natura loci*', Cic. *Leg. ag.* 2. 95).

This discussion, although seemingly a long way from Ovid's studied playfulness in the *Metamorphoses*, cuts to the heart of questions that are central to the intellectual background of Ovid's work. They find permanence and durability to be positive functions of a situational environmental ethos where Roman know-how is a product of natural epistemology and innovates by ripcut. Speaking against Rullus, Cicero was making his opening play in a consular career, transforming a man from nowhere (a *nouus homo*, an unknown quantity with no family lineage supporting his rise or protecting him from opposition) into a statesman and supreme authority. Cicero's metamorphosis (and relocation from the Romanised countryside, near Arpinum), like that of many other 'provincials' whose status and identities were being reinvented in these years, parallels and adds perspective to the threatened 'popular' transformation constructed so artfully and successfully as emerging from Rullus' proposition'. Given his highly personal stake in exploring how novelty and tradition can seamlessly unite (we will return to this when discussing grafting in case study 2), it is unsurprising that Cicero was interested in ideas of nature as generative and collaborative, but also dynamic, given some of his circling around Stoic pantheism.[24] This context also made him alert to the political sensitivities around the unnatural and monstrous, which he represents as the kinds of hybrid entity whose original defining qualities and parameters, marking its compound origins, persist, and contradict any superficial integrity.[25] This is in tune with his comment on Rullus' legislation that it overturns rational and unifying Roman practice ('*conuersa ratio*', *Leg. ag.* 2.68), practice which has brought Rome success and that can be conceived as a naturally ordered mode of existence tempered by ancestral custom (*mos maiorum*). Ostentatiously heterogeneous novelty would not be tolerated here.[26]

In *De Natura Deorum*, Cicero promotes natural metamorphoses as productive of wholly integrated and ordered entities; novel, but still subject to normal systems of progress. Environmental order and *ratio* therefore ensure stability but also structure and support naturally emerging transformation. Set alongside Lucretius' natural-science speculations in Book 2 of his epic poem, we can increasingly see how different genres and intellectual contexts are nonetheless evidencing a wider Roman concern to find in existence evidence of inherent dynamism but also internal consistency and unifying identity. Like Varro in his study of Latin, Lucretius' discussion of atomism uses the alphabet as an analogy to emphasise how stability and change are what produce complex, relative, and relational reality.[27] For Lucretius as for Cicero, ordered existence is only possible through a systematisation within which appropriate limits produce coherence, and incoherent hybridizations where the 'joins' remain evident are unsustainable.[28] These authors and their contemporaries were wary of things that looked like hybrids but also curious and determined to explain them into wholeness and a place within a pattern. As Lucretius puts it, what may look like individual elements (in language; in bodies) available for disaggregation and ripe for reconfiguration in new and previously unknown forms, are in fact already part of unities and only have their perceived meaning within those distinctive composite existing forms.[29] A monstrous hybrid, however, can only be a fantasy or a simulacrum because entities that are otherwise distinctively whole, without qualitative or qualitative interchangeability, are never viable when reconstituted from scratch in new forms. The results of these imaginative entanglements (Lucretius uses the example of a centaur) stir the imagination but can never embed themselves within the natural *material* and common-sense order of things.[30]

In this context, Lucretius' appetite for the wondrous and mysterious is significant. It demonstrates his interest in interrogating where material and metaphorical or imaginative realms intersect:

At conlectus aquae digitum non altior unum,
qui lapides inter sistit per strata uiarum,
despectum praebet sub terras impete tanto,
a terris quantum caeli patet altus hiatus,
nubila despicere et caelum ut uideare et aperta
corpora mirande sub terras abdita cernas.

But a pool of water no deeper than a finger,
which puddles between the stones on a paved street,
lays open downwards a prospect below the earth to such a reach
as from the earth the heavens' high vault gapes,
so that it seems as if one looks down upon clouds and heaven and sees exposed entities
marvellously secreted underground.

DRN 4.414–19

These lines suggest that semblance and reality can be powerfully connected on a material level but that wholeness and coherence are not always immediately self-evident to the human mind. They demonstrate how the perspectives made available through this system allow humans to become greater than themselves when recast as part of the truly dynamic and unified cosmos. Apparently disconnected entities or series of things, reimagined in a macrocosmic whole, manifest human artifice (a paved street) in an essential relationship with nature (how rain falls and is collected), and with a porosity in the experience and perception of the world around us. Within this model, what is truly real embraces all the possibilities produced by the fluid movement of atoms. Monstrosity rejects unity and highlights the joins; in the lines quoted above, a few drops of water show how nature inherently blurs boundaries and can help humankind to see a deeper, systemic, and participatory mutability.

Points of fusion and distinction are thus key to the ethical dilemmas posed by hybridisation, both in Cicero's thought and Lucretius' experimental cosmogony. I have suggested that these dilemmas became common currency for Roman intellectuals working on questions of identity and the impact of political and terratiorial transformation in this era, and their challenging perspectives on nature also underpin the more positive (or less threatening) ontological qualities of Ovid's representation of metamorphosis. This is therefore an argument that the creation of new things and their viability, rather than necessarily the process of material transformation, is the ultimate narrative of concern for those writing around the topic of metamorphosis in the late Republic.

Case study 2: Grafting (Varro, *De Re Rustica*)

Case study 1 explored the proposition that in a dynamic universe, change, at an atomic level and within ordered systems, is not about monsters or hybrid anomalies, nor reactive to incidental upheavals (which even the causality of Ovid's remarkable metamorphoses seems to presuppose). Instead, it is a crucial quality: operating for the most part invisibly to normal or untutored perception, and bound up in the creation of meaningful and viable entities and existential renewal. Without the right knowledge, however, the unexamined processes and outcomes of change are perilous. Varro's *De Re Rustica* returns us to the existential and community-defining qualities of land and 'stuff' explored in case study 1, but focuses on the manipulation of nature that enables scientific enquiry to support cutting-edge farming practice in propagation of crops. Like Ovid, Varro's attention to the hidden qualities of the forces that connect every stage in natural development, their powerful role in shaping human success, and the importance of scrutinizing each element in every transformation, makes clear how closely entwined plant and human lives and processes are.

Varro's introductory discussion of seeds and grafting (*Rust.* 1.39.1–40.6) is an important part of the narrative structure of Book 1, and his treatment is rather different to Virgil's lack of interest in seeds, and discussion of grafting as primarily a human *techne.*[31] In combining these two topics Varro showcases two complementary aspects of propagation as natural transformation, one (seeds) exemplifying the mysteries of cosmogonic forces and radical metamorphosis; the other (grafting) emphasizing the role of the (human) author. The crucial aspects are, first, the impossibility of perceiving the seeds of everything. Some transformations are not only difficult to explain but require a leap of imagination to conceptualise. This is expressed as a principle with reference to Anaxagoras (*Hist. Plant.* 3.1.4), and then followed up with the idea that everywhere within the air are invisible seeds (or primary elements of things) which are as complexly in motion as anything that is subject to air, and by terrestrial distribution water, as motile forces (*Rust.* 1.39.1). These motile forces are the cosmic beginnings of the environment, but from a practical perspective, visible seeds may be observed in their metamorphoses, and indeed '*id uidendum diligenter*' (they need to be kept a careful eye on, *Rust.* 1.39.1). Varro's contention here is important and connects lived experience directly to the scientific principles we saw in Lucretius; only through the scrutiny undertaken by landholders in the course of practical experiments in cultivation were seeds – and therefore a logic to natural transformations – first discovered ('*inuenit experientia coloni*', *Rust.* 1.39.2).

To undertake their metamorphosis and fulfil their potential for radical change in a way that satisfies this vision of a constructive Roman environmental understanding, seeds must be sown ('*neque, priusquam sata, nata*', *Rust.* 1.39.2). Where Varro takes this premise, however, is what most interestingly contextualizes Ovid. A little confusingly, but perhaps emphasizing his concern to focus attention on the radical quality of his cosmogony through the lens of agribusiness, Varro treats propagation through slips, or shoots and grafts, as forms of 'seed' growth. He explains this as the 'seed' or 'shoot' having wayward tendencies toward slipping out of the collaborative environmental balance: it does not 'intend' to transform from itself into a new and propagated feature of a well-run farm and will attempt its own divergent developmental process of budding and blossoming that will thwart the creator-farmer's agenda (*Rust.* 1.39.4). In effect, it has a divergent proposed life, a trajectory which the farmer must disrupt in order to form something different. This Varronian 'farmer' begins to look rather like the Ovid whose metamorphoses produce individual points of rupture to ensure that life and Roman history continue in broadly unifying story-arcs.

Pragmatically, Varro's farmer must tear from the main stock rather than simply amputating a branch, then reinvent the shoot by inhumation whilst its sap still flows.[32] The tear rather than a clean division is paradoxically what allows this new entity to become something whole and true to its authentic self. This model of propagation strengthens core identity between the old and new forms through a transformation predicated on ripping carefully but with a certain messy brutality to produce two new things each of whose generative potential reaffirms but also interrogates what replication of form means in practice. Same and different. The text's wider interest in category definitions and their socio-political significance for Roman territorial and agricultural success aligns this technical discussion contextually with assumptions and propositions relating to change and stability which also concerned Cicero and Lucretius (case study 1).

As we will see in case study 3, Catullus 64 also wove a story around the outcomes of unions of different sorts, with marriage and hybrid poetic form two clear signals of comparable concerns.[33] In Varro, the role of grafting is also a species of family history, similarly requiring shared and continuous narratives demonstrating kinship. Varro's interest in 'family' as a metaphor for change over time is not new to *De Re Rustica*, and is clearly significant in *De Lingua Latina* when he promotes the same-but-different qualities of grammar, syntax, and language in use.[34] That Varro's advice on grafting is not exceptional, in that any gardener knows it is correct, emphasises the significance of his decision (like Virgil's) to foreground grafting in a highly literary handbook on agriculture at a time when (based on the text's self- assertion as a very late work) Rome was emerging from civil war, recalibrating the concept of family and the body politic, and redefining core ideas of tradition and continuity.[35]

On grafting, just as much as when studying language, Varro emphasizes the need to understand the nature of the tree, the temporality, and the methodology (*Rust.* 1.40.5).[36] If we set this against Ovid's aetiological approach to metamorphosis we can also bring a new kind of focus to understanding Ovid's epic. Physical correspondence of some sort is crucial to the transformations in Ovid's poem as much as in Varro's thought; in Ovid there is also typically some element of the super- or hyper-natural, beyond scientific exactitude, involved – manifest in Varro's discussion of seeds. We may believe we have explanations for why transformations happen, but the how remains resistant. We see this too in Varro's emphasis on the waywardness of natural fertility. This is important here not just because of the way in which Varro describes the better success in transplanting a shoot broken from the stem (*deplanto*) than a broken-off (*defringo*) branch; he is describing something already with its own integrity (*Rust.* 1.40.4). It is important because when discussing grafts, he talks about a process whereby the shoot from one tree enters into a motile relationship with another ('*transit ex arbore in aliam*' *Rust.* 1.40.5) to produce something the same and different. The shoot is a disruptive force at the heart of growth; it represents new material emerging to its own scheme and signals the lurking potential for something radically different if one steps into a different episteme – just as

Ovid does when he materializes aetiological myth as a way of producing Roman truth about environmental and political forces.

Case study 3: Aetiology (Virgil, *Eclogue* 6 and Catullus 64)

Virgil, in introducing *Eclogue* 6, emphasises just how transformative an experience he wants his work to be (6.1–12). The *Eclogues*, a product of civil war, have been characterized as 'restless' and destabilising of equilibrium. In Virgil's song of Silenus, recounted by the shepherd Tityrus, these qualities are vividly present. Tityrus exists in the 'now', but his ability to slide out of time and into paradoxography, from Roman to mythic space, makes clear the poet's interest in intersections between time, materiality, identity and permanence; these are themes that return at scale in Ovid's *Metamorphoses*.[37] This final case study draws together many of the signals of 'metamorphosis' highlighted in case studies 1 and 2, instances that enable new ways of understanding what signs and structures might underpin a cultural sensitivity to metamorphosis before the Augustan programme recalled citizens to the tensions between evolution and revolution.

For Catullus (in poem 64) and Virgil, *Eclogue* 6, transformation is a precursor to knowledge, and knowledge emerges from aetiology in vivid, compelling form: stories that explain the hidden processes through which perceptual reality takes shape within historical time. The pursuit of causes was deeply woven into Roman literary culture of the first century BCE, and that scene embraced aetiology as a way to ground new hypotheses about Rome's future in studies of how and why events and practices came into being, or produced particular consequences. Catullus 64 is an important precursor to Ovid's dizzyingly nested episodic narratology because of the poem's extravagant deployment of ekphrasis (of the coverlet to the marriage bed of mortal Peleus and sea-deity Thetis, telling the story of Theseus' abandonment of Ariadne, 64.50–265) in a manner that destabilises the poem's frame. But the ekphrasis also centres transformation, and evaluates it as a lens through which to explain the past in such a way that it eventually supports a prophecy of the future. Catullus' coverlet cements a new identity for Thetis (obedient bride), it clothes the couch with a decorative scheme that undercuts the stability of the union it prefigures, and it connects those interwoven stories with the song of the Parcae through which they produce history as they weave.[38] Then the poem's final lines show how the cloth/text itself performs the act of metamorphosis: Ariadne's newly Roman guise, in which she would become something of a leitmotif across Ovid's varied output.[39]

Eclogue 6, like Catullus 64, is metamorphic, but depends on song for its most powerful message and most vividly realized metamorphoses, both centred on the character of Silenus. Unlike Catullus 64, it is explicit in extensively coding audience expectations around hodology: it self-declares as sprung from a ludic landscape in which experiment, and transformative projects, are built into its poetic positionality (*Ecl.* 6.1–2). Through lines 1–5 it flits between woods, battlefield and cultivated pastoral, developing a subtle *recusatio* before ostentatiously declaring this to have been a 'framing' structure and shifting to cosmology, creation stories and shapeshifting by way of an introduction to drunken old trickster Silenus (*Ecl.* 6.13–30), the poem's next singer.

A lushly corporeal drunkard, Silenus is also fashioned as a cosmic wordsmith in this version. Virgil's characterization will in this sense fall forwards into Ovid's staggering, sensual figure in ways that emphasise his significance as a touchstone for the entanglement between what is unseen or unspoken and what is real.[40] Silenus' song, as Brian W. Breed has observed, is a Virgilian 'phenomenon'.[41] This phenomenon is not simply echoing the kind of epic reported song his audience might have recalled from Apollonius' famous story of the Argo, presented by Catullus as a transformative and epochal moment in kickstarting what we might now term the anthropocene.[42] Captured Silenus' capture of Virgil's poem, and eventual characterisation as the cover-artist of Apollo's songs, clarifies what was implicit in Catullus: intertextuality and allusion as forms of metamorphosis.[43] Importantly for rethinking Ovid, Virgil's innovation here is his persistent emphasis on the storyteller's selective control over the playful and deceptive qualities of what appears to be tangible reality. Through this elasticity he gives Silenus the power to knit cosmic and humane modes intelligibly and seamlessly, despite the 'joins' (the chronological oddness of his organisation of the song's events). Looking from Catullus 64 to Virgil's creative and eclectic Silenus, and then on to Ovid's brief but expansively contextualised treatment of him in the *Metamorphoses*, we can see how each poet uses song as a formal agent and intuitive force through which the mutability of human reality and the fluid transience of the cosmos can meet.[44]

In Virgil's scheme worldbuilding is explicitly the focus, and the world he builds is not subject to the peaceful, slowed-down temporality of pastoral nor the *pax* extolled by Cicero (case study 1) as humanity's normative aspiration. *Eclogue* 6 encapsulates this in the juxtaposition of Silenus with another resonant figure: Orpheus. First, allusively within the poem, Orpheus' famed ability to make music a metamorphic agent (6.27–30). Apollonius' Orpheus already

> sang of how the earth, sky, and sea, at one time combined together in a single form, through deadly strife became separated each from the other; and of how the stars and moon and paths of the sun always keep their fixed place in the sky; and how the mountains arose; and how the echoing rivers with their nymphs and all the land animals came to be.[45]
>
> *Arg.* 1.496–502

Virgil's decision to develop the world's genesis as a product of Silenus' authoritative verses requires, but also offers, something different. To achieve this power for Silenus Virgil shows him first as a drunken pleasure-lover (*Ecl.* 6.13–26) so that the rupture between expectation and the transformation that he undergoes remains tangible through the twelve lines of cosmogony that intervene, before the tales change their nature and shift back into raunchier territory (*Ecl.* 6.43–63). Silenus both is and is not a changed character (Orpheus, but also Apollo, flickers in and out of his depiction), and this is also reflected in the language. Virgil first transforms Silenus by giving him a demiurgic voice, and with this shift comes a contextual ambiguity as he slips out of time in order to recalibrate and structure chronology from scratch (*Ecl.* 6.31–40). The sequence of metamorphoses starts as a concise cosmology but transforms into a series of sung vignettes characterizing metamorphosis as a companion of desire and then art (*Ecl.* 6.43–61, 62–73), before turning to the poisoned psyches and monstrous transformations associated with Scylla and Tereus (*Ecl.* 6.74–81).

The sequence of post-creation transformations that Silenus recounts is more than the sum of its parts. He begins with the world renewed after the flood and alludes to its repopulation by stones that metamorphose into a new race; next a cultural revolution: the beginning of civilization, Prometheus' theft of fire, homosociality (Hylas), and (by way of a perverse set of examples: the daughters of Proteus; Pasiphae) agriculture and thematically, the warping of desire in the struggle to (re)produce and maintain social bonds. Competition and ambition emerge by reference to Atalanta, while the Heliades, Phaethon's sisters (involuntarily transformed into signs of water, light and heat), bring together in summary three crucial elements for humankind to thrive.[46] Ovid, of course, would later entangle Atalanta and Orpheus (*Met.* 10.560–707) as an instance of the complex poetics of storytelling and dual identity, but the connection to her appearance as a sign of change in the cosmogony of *Eclogue* 6 has tended to be overlooked in favour of intertextual allusion linking her to Dido and Camilla.[47] Michael Paschalis sees dazzling Aegle, the naiad whose painting of Silenus' face (*Ecl.* 6.21–2) completed his transformation into a prisoner, as the thread linking 'the scientific and mythic accounts of the creation of the bucolic and agricultural landscape, of the generation of man and of the origin of fire'.[48] In my reading, Virgil's introduction of the Heliades (*Ecl.* 6.62–3) is also crucial backstory because their metamorphosis not only economically reintroduces fire (their family) and water (their fate) as topoi, it also presents the transformation as a process of overwhelming (*circumdare*) of one form (naming the Heliades as Phaethontiades) by another, to produce a new composite entity (sun and son), the water-loving alders.

Circling back to Catullus 64, the alder like the pine was a shipbuilding wood, and in this context the Heliades' presence in Virgil's sketchy history of humankind, drawing in Hylas (*Ecl.* 6.43–4), delicately recalls the opening to Catullus' poem and the adventurous destiny of Pelion's pines, soon to be transformed into the first ship (Catull. 64.10–11).[49] Virgil's bare-bones vanishing of Hylas makes the transformation situational – there, then not there (he was simply left behind, '*relictum*'), only hinting at the brutality and pathos that foreshadowed the loss of Heracles in Apollonius' telling.[50] Instead, the fountain and shoreline merge in the voices of the shouting sailors and the reverberations of their cries along the coast, forming a site of watery loss. This in turn might recall Catullus' Ariadne, but more immediately is echoed in elemental force of water signalled by the bark-clothed Sun's daughters. The transformations that Virgil develops by working to entangle words, themes, myths and literary allusions with such delicacy ensure that intertextuality between Catullus 64 and *Eclogue* 6 is almost too fleeting to capture, but cumulatively produces something radically intelligible in a Roman context where first principles were becoming increasingly sensitive, politically. It is this sense of an embodied song, present in the here-and-now but still understood as performed through the civilized monstrosity of more and less than human Silenus, that allows the shapes to shift before a reader's eyes.

Conclusions

We can be clear that Romans of the later centuries BCE did not automatically and comprehensively conceive the cosmos and all existing entities to be sentient or animate, nor to be suffused or vitalized by a divine energy or persona. Yet we can also be clear that Roman environmental thinking and understanding of the cosmos appreciated the role of the unknown, understood that a part of existence was played out by forces invisible and incomprehensible to most if not all humans, and that there were in some sense 'natural' laws which governed the world which (in effect) made possible attempts at heuristic understanding and scientific enquiry. In ancient Rome, as in other cultures before and since, when 'nature' is understood to exhibit 'sentience' and 'agency' these characteristics might be explained through some divine animating spirit, a form of anthropomorphism at heart, in at least some circumstances and for some 'sacred' entities. But it might also signal that something monstrous or threatening required further investigation and explanation in human terms.

In returning to Roman thought-experiments tackling environmentalism and the presence of multiple agencies within *natureculture* topographies, I have argued that what Tim Ingold calls 'processes of *growth*' are at the heart of 'human *concern*' with the environment and their/our place within it.[51] Such a weaving of growth, change and the nature of agency can be traced in Cicero, Lucretius, Varro and Virgil, through Ovid, to later and ostentatiously scientific works by Seneca the Younger and Pliny the Elder, and technical works by (for instance) Vitruvius, Columella and Frontinus.

It is not coincidental that it is Virgil's *Georgics* where Ovid's key verb *transformare* breaks cover.[52] My point has been, however, that each of these case studies makes clear the vitality of 'change' and in particular, change of form, in the imaginaries of the generations before Ovid. What we find in their colourful, richly sensory descriptions is an urge to speak to and from a world within which humanity and progress are part of a wider concept of developing order. Yet this is an order whose rules and boundaries, whose points of convergence between forms and elements, risk becoming monstrous in entities where their joins show. Virgil's vision, in this way, echoes motifs familiar from each of the authors I have highlighted, and clearly signals developing concern with boundaries and taxonomy and with what makes corporate entities function, emerging from the political and cultural fertility and turmoil of the first century BCE. If the idea of metamorphosis itself is an attempt to put superficial 'narrative' order onto a material reality which is inherently operating all at once and in every direction, this is a satisfying way of thinking more broadly about the notion of metamorphosis as much more conservative and counter-radical than might be imagined.

The reality of the pervasiveness, vitality and complexity of religious belief in antiquity (as now), as against power of tradition, ritual and cult practice, inevitably remains debated. Nevertheless, a desire to acknowledge and interrogate a quality of mystery underlying the materiality of change is evident in Ovid's intellectual hinterland. Rome's thinkers, natural scientists, and statesmen found themselves inhabiting a world in which almost everything was imagined to be more explicitly subject to forces of change than ever before, and the development of a better understanding of the visceral as well as the superficial characteristics and mechanics was one way in which a species of control might be asserted. Thus, literary works from classical antiquity, in Greek genres which Roman thinkers were beginning to appropriate and engineer into modes that could be expressed through Latin, were increasingly tackling these kinds of questions. Literary production of the late Republic characterizes a world in which the divide between human and non-human, mediated by a more or less pervasive sense of *numen* or 'ineffable force' at work in the cosmos, has many of the kinds of blurred lines that are only now, in the western European post-Enlightenment tradition, beginning to re-emerge in studies locating humanity as one among many agencies occupying and animating the universe.[53]

Ingold urges us to acknowledge that 'the forms of artefacts, like those of organisms, arise through processes of growth within fields of relationships' and 'artefacts grow like organisms, within the equivalent of a morphogenetic field'; in effect, everything 'is an original, not a replica'.[54] Ovid's complex narrative structure in the *Metamorphoses* works to expose and articulate the fallibility of a humanufactured epistemology which expects an intelligible and self-evident difference between original, replica, and hybrid. Any proposition of stability is dangerously illusory in the *Metamorphoses*, but as I have argued, this is not in itself a novel position when we read expansively across the ideas emerging from first-century BCE thinkers at Rome. Its originality lies in its willingness to test hybridity on multiple levels, situating itself amidst the perceived reality, traditional lore, and more or less creative speculation that environmental unpredictability produces. It does so within a particularly charged socio-political context in which ideas of same and different, custom and revolution, have taken on bloody and fatal consequences and emerged in new forms. There was a discomfort with this novelty which Augustus' own testimony, delivered through his statement of achievements (*Res Gestae*), tacitly acknowledged, and which is also a factor in Ovid's distinctive response.

By contrast to the strategies adopted by earlier authors, Ovid's metamorphic environmental vision makes no promises or predictions for the wider relationship between transformation and stability within Rome's relational landscape writ large. His exposition of multiply infolded individual voices, entities and agencies reflects and engages with previous generations' politicization of the environment as a metaphor, but instead of finding the potential for harmony and familiar order operating across seasonal time through a linear or regularly cyclical sequence, he offers instead a world in which *neither* the expected *nor* the surprising is indelibly or persistently 'before' or 'after' the other.[55] A world where monstrosities from the ancestral psyche can draw life from the aetiological impetus to articulate the emergence of boundaries within *natureculture* and civil space. A story, therefore, of a world where all those once-upon-a-time stories come true.

Notes

1. For the numbers, see Neuru 1980: 3. See Lafaye 1904 for the Greek background to metamorphosis in myth.
2. Haraway 2003; see the overview and discussion by Latimer and Miele 2013.
3. In the wake of Gibson 1977: 67, defining 'affordances' as 'a specific combination of the properties of [a thing's] substance and its surfaces taken with reference to an animal', the question of human exceptionality, and the relativity of human experience to that of other 'animate' species, has been Haraway's focus, developing models of enormous value for understanding the intellectual landscape from which *Metamorphoses* emerged; Haraway 2008 in particular, but also Haraway 2006.
4. Although the evidence is thin, Parthenius seems to have been brought to Italy from Bithynia if not enslaved, then at least as a captive (PoW), and to have gained standing in a literary coterie including Virgil. On the text of Parthenius *Metamorphoses*, Lightfoot 1999; on his contextual significance, Francese 2001.
5. Especially Ingold 1993, 2011, but also Ingold 2006, 2012 and e.g. Schilhab 2017. On the perception of time, the ever-polemical Tallis 2017 is worth exploring.
6. Abram 1997 (first published 1996).
7. My working hypothesis on what constitutes 'the body' for this paper draws significantly on Holmes 2010: 64–70.
8. Bernard 2018: 16.
9. We see evidence of the concern with 'family' and innate qualities in e.g. Varro's interweaving of discussion of *genus* through his politically engaged study of Latin (references at Spencer 2019: 193–94 and at n.42).
10. See e.g. Rüpke 2007. See also Myers' observation: 'plants have incredible sensory dexterities, and [. . .] they can make sense of and actively intervene in their worlds', Myers 2015: 36. On the ancient nuances connecting and reframing divine and human agencies, politics, and ecology, see Lane 2019, and (in a sober polemic against what she terms 'the modern myth' of pagan animism and the challenge in translating *numen*) Hunt 2019.
11. Based on internal references, the traditionally accepted fixed dates situate the poem's composition from 36 to 29 BCE (see the helpful list of internal references in Horsfall 2001: 93).
12. All translations are my own, except where specified otherwise. The Latin texts used throughout are those from the Loeb Classical Library series unless otherwise specified.
13. This deployment of shifting perspectives to destabilise any unthinking assumptions about what makes land *Roman*, and how this shores up the appearance and reality of Roman identify and existence, is evident throughout, but already in place at e.g. Verg. *Georg*. 1. 56–9 where Roman land can be simultaneously recognizably full of traditional crops and packed with the marvels of empire.
14. On ancient approaches to environmental determinism, see Kennedy and Jones-Lewis 2016.
15. The *ager publicus* was territory gained through Rome's expansion, and 'publicly' owned, but whose disposition was polarised between the suspicion that it was either monopolized on preferential terms by wealthy leaseholders (which seems to have been especially the case in, e.g., Southern Italy), or the entitlement of poorer Romans through redistribution. Legislation from 367 BCE set a limit of 500 *iugera* to any individual's holding. It is generally accepted (though poorly documented) that much of this land was in practice held by Latins and allied peoples, and growing demand for this land by the late second century BCE had already

produced contentious legislative activity, leading to violence, as part of the Gracchan project to 'privatize' (grant more security of tenure and ownership to) *ager publicus* to strengthen the rights of smallholder occupiers. For more on the detail, see Roselaar 2010.

16. On the development of *pax* and Pax in the late Republic, see Weinstock 1960.

17. Harper 2019: 14, and on universal empire, e.g. Verg. *Aen.* 1.278–9, Lucr. *DRN* 2.90–4, Ov. *Fast.* 2.684.

18. Harper 2019: 20–1, and passim.

19. '*Age, non definis locum; quid? Naturam agri?*' (Cic. *Leg. ag.* 2.67). Cf. Varro, *Ling.* 6.56 connecting *locus* through direct etymological relationship with *loqui* (to speak) when defining what constitutes the Latin language.

20. Forni 2019 provides a detailed overview of how some of this 'collaboration' might emerge from the evidence.

21. '*Alterum genus agrorum propter sterilitatem incultum, propter pestilentiam uastum atque desertum emetur ab iis, qui eos uident sibi esse, si non uendiderint, relinquendos*' (Cic. *Leg. ag.* 2.70).

22. Cordovana and Chiai 2017: 12–13; relevant on ethos and ambience, in the same volume, e.g. Marcone 2017, Nelis-Clément 2017. On the changing Roman perceptions of the 'value' and purpose of land in a globalized economy, Marcone 2019.

23. '*Hoc enim est usus, quasi de aliqua sentina ac non de optimorum ciuium genere loqueretur*', for this is the expression [Rullus] used, as if he were speaking of liquid waste and not of a class of excellent citizens (Cic. *Leg. ag.* 2.70). For *sentina*, e.g. Verg. *Georg.* 3.309; Liv. 1.59.9.

24. Cic. *Nat. D* 2.58, 83–4.

25. Cic. *Nat.D* 2.5, 29. Compare Cicero on Verres as a *monstrum*, combining the worst of pig (*uerrinus*) and broom (*euerriculum*) in his destructivity (*Verr.* 4.47, 53).

26. For another version of this position, see e.g. Varro, *Ling.* 9.6.

27. Lucr. *DRN* 2.695–706. Using language as an analogy, Varro suggests that to recognize and integrate the best in novel and potentially exemplary practice requires, in the radically changing times that he inhabited, a new epistemic toolkit (*Ling.* 9.6, 18), on which Spencer 2019: 37–8.

28. See especially Lucr. *DRN* 2.700–6.

29. 'I do not hold that there are very few common letters running through all [words] or that no two words, compared, are made up of the same letters, but that it is uncommon for them to be so alike' (Lucr. *DRN* 2.692–4).

30. Lucr. *DRN* 2.739–48.

31. Contextualizing *Georgics* 2.69–82 on grafting, broadly, Lowe 2010.

32. For this method see also e.g., Verg. *Georg.* 2.23.

33. On the metapoetics of genre and intertextuality in Verg. *Georg.* 2, another angle on grafting hybridity, see Henkel 2014.

34. See Spencer 2019: 38–9, 42–5, 49–50, 71–2, 91–3, 182–3, 184–213 (*passim*).

35. See also Var. *Rust.* 1.40.6 on the 'new' mode of grafting for the violence inherent in the process.

36. Cf. Var. *Ling.* 5.13; 7.4 (cf. *Ling.* 7.9; 5.74).

37. Seider 2016, 3. See also in particular the expansive and important discussion by Paschalis 2001 which has been especially influential for my reading. For the atmosphere of trickery

and 'play' developed through the opening 'frame', Verg. *Ecl.* 6.1, 19, 23, 28. It is not novel to suggest that the cosmogonies of Lucretius and Virgil are the backdrop for Ovid's version in *Met.* 1 (e.g. Wheeler 1995); rather, I suggest that these earlier impulses towards understanding existence as metamorphosis form part of a broader sense, in the first century BCE, that chaos, pluralism and multimodal perspectives are what give dimension and persistence to reality.

38. Catull. 64.323–81. See Laird 1993: 28–9. Catull. 64.405 brings time to the 'now'. On temporality and the woven coverlet as a kind of time-machine, Wasdin 2017. Cf. Parcae at Verg. *Ecl.* 4.46–7.

39. Scheid and Svenbro 1996: 136. In *Heroides* 10 and *Ars Amatoria* 1, Ovid is 'retelling' the same stages in Ariadne's 'story' as Catullus 64, and *Heroides* 10.96–8 clearly evokes Catull. 64.152–3 (as Armstrong 2006: 223 observes). Catullus 64, as background to Ovid's Ariadne, emphasises that working intertextually is also a process of metamorphosis. *Fast.* 3.459–516 is Ovid's third significant engagement with Ariadne (where she becomes the Roman Libera) and moves the story past Catullus. Ariadne does feature briefly in the *Metamorphoses*, 8.172–82 (where the metamorphosis featured is the straightforward one of her crown). Ov. *Met.* 6.1–145 is where weaving breaks cover most explicitly (the weaving competition between Arachne and Minerva), allowing Ovid to produce two 'competing' versions of what showstopping work that ekphrasis and 'text' can achieve.

40. Ov. *Met.* 11.89–105, where Silenus becomes the trigger for Midas' ill-judged request for the 'golden touch', but more broadly, his series of tragi-comic interventions in disturbing rape scenes in the *Fasti*.

41. Breed 2000: 328.

42. Catull. 64.11: the Argo's first voyage was a rite of passage for Amphitrite, a new experience for a newly experienced sea. Compare Ap. Rhod. *Argon* 1.496–511, on which see Breed 2000: 330.

43. On intertextuality as metamorphosis see Peirano 2009 (Virgil's Scylla). Davies 2004 is crucial for understanding the motif of binding for my reading of Silenus. On Silenus as an Alexandrian, see Skutsch 1956: 193–4.

44. This reading picks up on an interpretation proposed by Segal 1969a: 418.

45. Translation: Race 2008.

46. On the complexities of why Virgil calls them Phaethontiades, and how this further entangles Apollo as creator and destroyer, I find Huyck 1987 to be convincing.

47. See e.g. Ciabaton 2020.

48. Paschalis 2001: 218.

49. For the alder, Verg. *Georg.* 2.451. Of course, Catullus' opening to poem 64 itself evokes Ap. Rhod. 1 in multiple complex ways.

50. Ap. Rhod. *Arg.* 1.228–33.

51. Ingold 2000: 77, 76, original emphases.

52. Verg. *Georg.* 4.440–2; followed by Verg. *Aen.* 7.416.

53. E.g. Bennett 2004. See also Iovino and Oppermann 2014, in conjunction with their growing body of scholarship e.g. Iovino 2016; Iovino and Oppermann 2012; Oppermann and Iovino 2017.

54. Ingold 2000: 291, 372.

55. This reading complements the conclusions reached by Hardie 1995 on the Empedoclean speech of Pythagoras (Ov. *Met.* 15). It also echoes my conclusions on the significance of Catullus' coverlet in this literary backstory.

CHAPTER 10
'WHO CAN IMPRESS THE FOREST?' AGRICULTURE, WARFARE AND THEATRICAL EXPERIENCE IN OVID AND SHAKESPEARE

Sandra Fluhrer

Human impact on global warming and the destruction of ecosystems raises urgent questions about the possibility of historical learning. How can we bring about the necessary changes in human behaviour towards the earth? How can we approach past, present and future events related to the Anthropocene as learning opportunities? As an aesthetic form with transhistorical traits mythological literature can make important contributions to our thinking about these questions. If we consider mythological texts (alongside other art forms) as forms of *experience* – understood in Bernhard Waldenfels' terms as poetic transformations of bodily affect or pathos that provoke an equivalent form of response in readers[1] – their analysis is not only 'work' on the collective imaginations that allow the myths to develop and evolve.[2] It may also become work on the experiential foundations of myths and their reception. Mythological literature triggers aesthetic experiences, which implicate the bodies and minds of its readers in multifaceted ways. Recent research on aesthetic experience and 'somaesthetics' allows us to understand that the body is an important site of response for readers and viewers of art.[3] Studies of body memory, for instance, imply that past and present experiences are viscerally entangled in our reading of mythological literature.[4]

Within this conceptual horizon, this chapter investigates the relation between warfare, agriculture and aesthetic experience in the Cadmus episode from Ovid's *Metamorphoses* (c. 8 CE) and in Shakespeare's *Macbeth* (1606). In recent decades, the impact of agriculture on climate change has become increasingly clear.[5] This insight is not yet present in Ovid and Shakespeare; however, the intricate connections between agriculture and warfare which their texts and relevant intertexts exhibit suggest poignant ambivalences in how Classical Roman as well as Elizabethan and Jacobean societies might have approached anthropogenic processes of cultivation as destruction. Both Ovid and Shakespeare foreground the role of somatic and sensory experience in their representation of the complex interlacings of agriculture and warfare. Their texts are marked by an extraordinary poetic plasticity, an art of creating lively, tangible characters and scenes which contribute to intense aesthetic experiences in their readers and audiences.[6] These are not only intellectual, but also visceral experiences that originate in tactility and kinaesthetic feeling – in the sense of touch which, since Aristotle (*De anima*, 423a–4a, 424b), has been regarded as encompassing all senses through a sensory network.[7] In this chapter, I seek to demonstrate the profound analogies that appear between Ovid's and Shakespeare's poetic strategies for conveying, processing and connecting historical, contemporary and possibly future experiences around links between agriculture and warfare. For these strategies, forms of theatricality prove to be crucial for Ovid, as well as for Shakespeare.

I

A prerequisite for both agricultural terrain and forest, land and soil provide a necessary means for human subsistence on earth. Land is also related to ideas of the political, such as cultivating and founding, and to political actions such as defending or expanding territory. For politics, land and forest have a mythic or representational and a material function; they are used as sources for imagery of the origins and development of political communities, but also as concrete resources to physically build, change and expand territory, between cultivation and conquest.

In Classical Greece, owning land was a prerequisite for politics: most citizens and thus also most soldiers were farmers. In his study *Warfare and Agriculture in Classical Greece*, Victor Davis Hanson writes that 'war was endemic, and the energies of the citizens were largely consumed with either working, protecting, or attacking cropland'.[8] The destruction of grain was an integral part of warfare.[9] 'Almost every consideration of a Greek army – logistical, tactical, strategic, psychological, and technological – was in some way connected to agriculture,' Hanson states.[10] Adrian Goldsworthy adds that hoplites were mostly small farmers and thus hoplite warfare obtained 'a peculiar rhythm of its own, fitting in with the agricultural year'.[11]

In Roman Republican and early Imperial times, the relation between warfare and agriculture had changed but continued to be a crucial issue. As Dominik Maschek explains, the distribution of conquered land to Roman citizens had always been a concern in Roman politics, and became even more so after the civil wars between 44 and 30 BCE.[12] In Augustan times, the veterans were given land; the processes of expropriation and redistribution often entailed violent conflict,[13] traces of which appear in Augustan literature, in the opening dialogue of Virgil's *Eclogues*, for instance (1.11–13, 64–83).[14] As Maschek holds, around 700,000 people profited from land redistributions during the triumvirates, and, if we can believe Augustus' *Res gestae*, he resettled or gave land to 300,000 veterans during his reign.[15] The plough in Augustan iconography became a symbol of colonization; inadvertently, it also epitomized expropriation and ruin.[16]

Towards the end of the first book of Virgil's poem on agriculture, the *Georgics*, a dark prophecy is pronounced: *agricola incurvo terram molitus aratro / exesa inveniet scabra robigine pila / aut gravibus rastris galeas pulsabit inanis / grandiaque effossis mirabitur ossa sepulcris*. – 'the farmer drudging soil with his curved plough / will turn up scabrous spears corroded by rust / or with his heavy hoe strike empty helmets, / and gape at massive bones in upturned graves' (1.494–7).[17] Virgil's verses point at the ambivalent functions that land holds between farming acre and battlefield, at the multiple layers of history that haunt the present, and at natural processes of metamorphosis within the soil, which continuously change its composition.

More markedly than Virgil, to whom both are indebted, Ovid and Shakespeare intertwine mythic narratives of metamorphosis with contemporary discourses of natural history and politics in their depiction of war. Ovid takes the Cadmus story beyond its mythological context, recalling connections between warfare and agriculture in Roman imperialism. While in *Macbeth* mythical metamorphosis turns into military camouflage, the play echoes mythological imagery as well as ecological issues of the Elizabethan age, such as deforestation. As I will show, Ovid and Shakespeare blur the lines between mythical metamorphosis, poetic metaphor and the materiality of both landscape and war. When stable ideas of land begin to shift in the processes of aesthetic experience that the texts initiate, theatre comes to the fore as an always provisional and experimental medium that appears to be especially suited to processing experiences of ambivalence and contradiction.

II

Ovid narrates the story of Cadmus in two parts. The first part opens Book 3 of the *Metamorphoses*: Cadmus is exiled by his father, Agenor, and given the task of searching for his sister Europa, who had been abducted by Jupiter in the shape of a bull. Unable to find his sister, Cadmus turns to Apollo's oracle for advice on where to live. He is told to found a city in a spot designated by a cow: the Boeotian city of Thebes. Before he can build the city, Cadmus must defeat a huge snake related to the god of war, sow its teeth into the earth, witness a people of warriors grow from the furrows, and the immediate outbreak of civil war between them. Only five of the earth-grown men survive and help found the city, which, as is well known, harbours a number of unhappy destinies. Some of these calamities lead Cadmus, late in his life, to ask whether the snake he slew had been sacred. Fulfilling a prophecy he received after the killing, he then turns into a snake himself.

This episode's treatment of the relation between agriculture and warfare has received little scholarly attention. Virgil's *Georgics* must have provided especially fertile inspiration for Ovid's interest in the Cadmus myth's significance as a repository for agricultural concerns, both symbolic and historical. It contains the myth of the dragon teeth (cf. 2.140–8), a battle with a snake (3.416–439), as well as concrete and metaphorical links between warfare and agriculture (1.490–514, 2.458–60, 3.339–48, 4.67–90). As I seek to show, Ovid's version of the Cadmus myth leaves open the question of whether dangers for civilization arise from (supposedly) untamed nature, or from processes of cultivation. In the undercurrents of his Cadmus episode, he asks for an art of human cultivation that does not amount to raising armies.

At the beginning of the episode, Apollo describes the cow, which by lying down provides the sign for the place of the city's foundation, as being without any trace of the plough.

> vix bene Castalio Cadmus descenderat antro,
> incustoditam lente videt ire iuvencam
> nullum servitii signum cervice gerentem;

subsequitur pressoque legit vestigia passu
auctoremque viae Phoebum taciturnus adorat.
iam vada Cephisi Panopesque evaserat arva;
bos stetit et tollens speciosam cornibus altis
ad caelum frontem mugitibus impulit auras
atque ita respiciens comites sua terga sequentes
procubuit teneraque latus submisit in herba.
Cadmus agit grates peregrinaeque oscula terrae
figit et ignotos montes agrosque salutat.

Cadmus left the holy cave
And saw, almost at once, as he went down,
A heifer ambling loose that bore no sign
Of service on her neck. He followed her
With slow and wary steps and silently
Worshipped Apollo, guardian and guide.
Now past Cephisus' shallows and the meads
Of Panope they wandered on, and there
The heifer stopped and raised towards the sky
Her graceful high-horned head and filled the air
With lowings; then, her big eyes looking back
Upon her followers, she bent her knees
And settled on her side on the soft grass.
Cadmus gave thanks and kissed the foreign soil,
Hailing the unknown hills and countryside.

Met. 3.14–25[18]

The verses recall the description of the golden age and the conception of the ages of the world as a progressive decay and degeneration in the first book of the *Metamorphoses* (as well as in the *Georgics*[19]), where the earth provides crops without being ploughed (*Met.* 1.101–2, 109–10). However, subtle traces of cultivation disturb the impression of a foundation on untouched land. When Cadmus encounters the cow, three consecutive verses provide three expressions of 'earth'. The degree of culturalization increases with the different terms. The cow lays down *in herba*, which denotes meadow for grazing but also plants that are useless to humans. Cadmus fastens kisses (*oscula [. . .] figit*) on this ground, which is now called *peregrina terra* ('foreign earth') – vocabulary of colonization and Roman othering. The third expression explicitly evokes agriculture: Cadmus salutes the unfamiliar mountains and fields, *agros* (*ager* is also covertly present in *peregrina*). Roughly 80 verses later, Cadmus will become an 'agriculturalist':[20] He ploughs the ground and sows the snake's teeth from which civil war is to emerge. But verses 24–5, which introduce the *peregrina terra*, cast doubt on the pristine nature of the soil, and cause us to question the colonial myth of untouched land. Hinting at Roman imperialism, they point at what Robert Harrison described as 'the insatiable mouth of empire devour[ing] land, clearing it for agriculture and leading to irreversible erosion in regions that were once the most fertile in the world'.[21]

After his first contact with the land, Cadmus has his servants fetch libation water from a running well, a task from which they do not return, falling victim to the snake:

> sacra Iovi facturus erat; iubet ire ministros
> et petere e vivis libandas fontibus undas.
> Silva vetus stabat nulla violata securi
> et specus in medio virgis ac vimine densus
> efficiens humilem lapidum compagibus arcum,
> uberibus fecundus aquis, ubi conditus antro
> Martius anguis erat, cristis praesignis et auro;
>
> *He sent his henchmen forth to find a spring*
> *Of living water for the ritual.*
> *There stood an ancient forest undefiled*
> *By axe or saw, and in its heart a cave*
> *Close-veiled in boughs and creepers, with its rocks*
> *Joined in a shallow arch, and gushing out*
> *A wealth of water. Hidden in the cave*
> *There dwelt a snake, a snake of Mars. Its crest*
> *Shone gleaming gold;*
>
> *Met.* 3.26–32

Again, the idea of a *locus amoenus* is installed – and quickly subverted: The forest is old (*vetus*), and thus capable of bearing history; the curved grotto, *specus*, which hosts the dragon-like snake, prepares for the theatre imagery to come, and oscillates between nature and culture; *virgae* and *vimen* are weaving material; the stone wall is manufactured: *efficiens*. When Cadmus eventually enters the forest and fights the beast, his first weapon is a *molaris* (3.59), which dictionaries list as a poetic term for a large stone, but which in its first sense is a millstone, and thus a cultural product. In line with this meaning, Ovid notes that the impact of the stone would have shaken steep walls with high towers – but not the snake (61–2). Its defence is described in military terms: *loricaeque modo squamis defensus et atrae / duritia pellis validos cute reppulit ictus* ('its scales like armour shielding it, / Stood fast unscathed; its hard black carapace / Bounced the blow back', 63–4). Thus, the division of nature and culture is disturbed even before the foundation of the city whose descendants are to 'become the victims of a natural world hostile to man'.[22]

The episode's interlacing of agriculture and warfare is most prominent in the description of the Spartoi, the warriors growing from the dragon teeth that Athena has Cadmus sow into the furrows. These dental 'seeds', booty from the fight with the snake, produce warriors as crops, who immediately start a civil war, before the survivors found the war-struck city of Thebes.

inde (fide maius!) glaebae coepere moveri,
primaque de sulcis acies apparuit hastae,
tegmina mox capitum picto nutantia cono,
mox umeri pectusque onerataque bracchia telis
exsistunt, crescitque seges clipeata virorum.
sic, ubi tolluntur festis aulaea theatris,
surgere signa solent primumque ostendere vultus,
cetera paulatim, placidoque educta tenore
tota patent imoque pedes in margine ponunt.

The tilth (beyond belief!) began to stir:
First from the furrows points of spears were seen,
Next helmets, bright with nodding painted plumes,
Then shoulders, chests and weapon-laden arms
Arose, a growing crop of men in mail.
So, when the curtain at a theatre
Is raised, figures rise up, their faces first,
Then gradually the rest, until at last,
Drawn slowly, smoothly up, they stand revealed
Complete, their feet placed on the fringe below.

Met. 3.101–23

Astonishingly, Ovid compares the rise of the soldiers to a rising theatre curtain (111–14). As his verses show, Roman theatre curtains were drawn up from the ground at the end of a performance, gradually revealing their embroidery.[23] There is an evident connection between the anachronistic simile of the theatre curtain and Roman imperialism. In the *Georgics*, in a passage which is concerned with the project of building a temple for Octavian, Virgil mentions a specific curtain embroidered with Britannic warriors: *iam nunc sollemnis ducere pompas / ad delubra iuvat caesosque videre iuvencos, / vel scaena ut versis discedat frontibus utque / purpurea intexti tollant aulaea Britanni.* – 'Even now I'd joy to lead the ritual / procession to the sanctuary, to watch the bullocks' sacrifice, / or see the stage-scene vanish when the sets are turned, / how its embroidered Britons lift the purple curtain' (3.22–5).[24] The Roman conquest of Britain is yet to come. However, since a list of conquered regions follows the verses about the curtain, the poem connects these verses to imperialism and turns them into an Augustan vision of conquest. The theatrical performance the verses imagine are part of a religious-political festival.

Roman theatre is considered by and large a medium for political publicity and influence, rather than a means for negotiating the polis as in Classical Greece.[25] This must have been even more the case for amphitheatres and their *spectaculi*, whose origin and development was closely related to the 'Roman military ethos of the middle and late Republic', as Katherine Welch holds.[26] Welch, and more recently also Maschek, argue for a connection between the construction of amphitheatre buildings and veteran colonization.[27] Being built primarily as instruments of social control (concerning seating rules for instance), new amphitheatres in the colonized areas inadvertently must have also worked as platforms for processing recent colonial experiences.[28] While games in amphitheatres differ markedly from dramatic performances in regular theatres in terms of the role the actors' bodies play, the two cultural practices were not strictly separated in Rome; they shared similar themes and aesthetic principles, and they often took place side by side and in the same venues. Regular theatres were sometimes used for gladiatorial games,[29] a variety of theatrical practices and arena games belonged to 'the eclectic atmosphere of Roman theatres', and mime acts were often part of gladiatorial games.[30] Rome's Campus Martius could be seen as a place of uncanny re-enactment: being alternately used as rangeland and for military exercise, it increasingly became a place of imperial representation, among others with its (amphi-)theatre buildings, such as the Theatre of Pompey (completed in 55 BCE) as well as three Augustan venues: the Amphitheatre of Statilius Taurus (30 BCE), Rome's first stone amphitheatre, and the Theatres of Marcellus and Balbus (13 BCE).[31]

Ovid's curtain image, which takes up elements of Virgil's vocabulary in the verse *sic, ubi tolluntur festis aulaea theatris*, is more than a casual reference. It addresses the relation of war experience and aesthetics and it invites readers to consider how different media transform the experience of war. While the older Augustan poets experienced the civil war directly, Ovid only witnessed its 'afterlife' in speeches, poetry, visual arts and theatre. The curtain image reinforces our sense of his mediated experience of civil war. Roman theatre curtains contributed to strategies of audience impression and enchantment.[32] Ovid's curtain comparison addresses the relationship between plasticity and experience. Comparing the growing warriors to images on a theatre curtain, Ovid initially flattens the representation. However, as soon as he has established the still image, he enlivens it again by using direct speech and describing drastic physical action which could be part of a bloody amphitheatrical spectacle rather than a representational performance. Ovid uses the image of the frozen and flat bodies on the curtain to make the following war scene the more tangible. This quick confrontation of two aesthetic modes invokes an opposition between the flatness of Augustan aesthetics of representation and Ovid's own vibrant and multisensory verse. While the former tends to mythicize and mystify war experience, Ovid's poetry pledges for an art of multisensory experience and an intricate play with involvement and distancing – a sensual form of theatricality.[33]

Theatrical and amphitheatrical aesthetics are interlaced in this episode. This interweaving brings into question the distinction between physically harmless representational performance and fatal fighting games and thus, in fact, represents the fluidity of Roman *ludi*.[34] That Ovid combines these (nevertheless) heterogeneous cultural practices within his text is provocative, but it also celebrates the potential of poetry to evoke intense experiences without causing physical harm.

The medium of theatre belongs to the literary history of the Cadmus theme.[35] Ovid enhances the connection to theatre not only in the curtain image but throughout the whole episode. The grotto in the forest has a theatrical shape. The sequence of fights alludes to the course of events in a Roman amphitheatre, with beast fights in the morning, followed often by executions, and gladiatorial fights in the afternoon:[36] the fatal encounter of Cadmus' men with the beast, his fight against the snake, finally the brief but bloody civil war scene. Ovid reports that Cadmus fights the snake after noon (*Fecerat exiguas iam sol altissimus umbras*; / 'The noonday sun had drawn the shadows small,' 3.50). All three fighting scenes could be narrated by a spectator with a broad and slightly elevated perspective.

The episode abounds with references to looking and watching (as does the third book as a whole[37]). The most striking of such instances is the prophecy of Cadmus' transformation:

> Dum spatium victor victi considerat hostis,
> vox subito audita est (neque erat cognoscere promptum,
> unde, sed audita est): 'quid, Agenore nate, peremptum
> serpentem spectas? et tu spectabere serpens.'
> ille diu pavidus pariter cum mente colorem
> perdiderat, gelidoque comae terrore rigebant;
>
> *Then as the victor contemplates his foe,*
> *His vanquished foe so vast, a sudden voice*
> *Is heard, its source not readily discerned,*
> *But heard for very sure: 'Why, Cadmus, why*
> *Stare at the snake you've slain? You too shall be*
> *A snake and stared at.' For an age he stood*
> *Rigid, frozen in fear, his hair on end,*
> *His colour and his courage drained away.*
>
> *Met.* 3.95–100

Ovid artfully entangles snake and spectatorship. *Draco*, one of the episode's other words for snake, is related to the Greek verb *dérkomai* (see). Yet, the passage also addresses hearing and a wide concept of touch, encompassing tactility, feeling, and kinesis. In the verse *serpentem spectas? et tu spectabere serpens* (*Met.* 3.98) all syllables are involved in alliteration and/or assonance making it almost become a tongue-twister and reminding of the hissing and rattling of snakes. The chiastic structure imitates snake movements. The passage surpasses classical conventions of alliteration, assonance and syntax in Latin poetry pushing language almost beyond a means of human articulation, thus foreshadowing Cadmus's destiny not only lexically, but also in a vibrating form of body language.

In the fourth book of the *Metamorphoses*, Ovid returns to Cadmus. Disillusioned by the many misfortunes of his family and the city of Thebes, Cadmus leaves the city. After long wandering he and his wife Harmonia arrive at Illyria, old and bent. Cadmus asks whether he had slain a sacred snake and begs to become a snake himself, whereupon he is transformed:

> dixit et, ut serpens, in longam tenditur alvum
> durataeque cuti squamas increscere sentit

nigraque caeruleis variari corpora guttis;
in pectusque cadit pronus, commissaque in unum
paulatim tereti tenuantur acumine crura.
bracchia iam restant; quae restant bracchia tendit,
et lacrimis per adhuc humana fluentibus ora
'accede, o coniunx, accede, miserrima' dixit,
'dumque aliquid superest de me, me tange manumque
accipe, dum manus est, dum non totum occupat anguis!'

Even as he spoke he was a snake that stretched
Along the ground. Over his coarsened skin
He felt scales form and bluish markings spot
His blackened body. Prone upon his breast
He fell; his legs were joined, and gradually
They tapered to a long smooth pointed tail.
He still had arms; the arms he had he stretched,
And, as his tears poured down still human cheeks,
'Come, darling wife!' he cried, 'my poor, poor wife!
Touch me, while something still is left of me,
And take my hand while there's a hand to take,
Before the whole of me becomes a snake.'

Met. 4.576–85

Ovid narrates this metamorphosis in great detail with attention to the role of the senses, touch in particular. First Cadmus' skin hardens, gets scales, and changes its colour; then he loses the ability to walk upright. Arms and face remain a little longer. Finally, he loses his voice. Harmonia laments his loss of form, before the two rejoin, developing new forms of touch and contact:

dixerat. ille suae lambebat coniugis ora
inque sinus caros, veluti cognosceret, ibat
et dabat amplexus adsuetaque colla petebat.
quisquis adest (aderant comites), terretur; at illa
lubrica permulcet cristati colla draconis,
et subito duo sunt iunctoque volumine serpunt,
donec in appositi nemoris subiere latebras.
nunc quoque nec fugiunt hominem nec vulnere laedunt,
quidque prius fuerint, placidi meminere dracones.

He licked his poor wife's cheeks, and glided down
To her dear breasts, as if familiar there,
And coiled, embracing, round the neck he knew.
All who were there – and courtiers were there –
Were terrified; but she caressed and stroked

Her crested dragon's long sleek neck, and then
Suddenly there were two, their coils entwined.
They crawled for cover to a copse nearby;
And still, what they once were, they keep in mind,
Quiet snakes, that neither shun nor harm mankind.

Met. 4. 595–603

Harmonia tunes in; she gently strokes the snake's slippery neck and eventually becomes a snake, too.

Only at the very ending of the passage does Ovid reveal that Cadmus' metamorphosis takes place in front of an audience. It must have seemed important to Ovid to postpone this realization. Readers or hearers of the episode are supposed to experience the metamorphosis and gentle snake life (and love) on their skins, almost as if they become snakes themselves. Having established this impression, Ovid suddenly turns the scene into theatre. This is the moment when we as readers realize our physical involvement. After mentioning the audience, Ovid creates a verse whose mimetic quality piercingly expresses Harmonia's caressing: *lubrica permulcet cristati colla draconis*. The alliterations come close to eliciting the goose skin which had befallen Cadmus when he received the prophecy of his metamorphosis. The verse marks the difference between man and snake, but at the same time demonstrates to the readers their physical, their experiential involvement in the process of metamorphosis that they have witnessed.

Cadmus' and Harmonia's snake life might be what comes closest to a happy ending in the *Metamorphoses*. In Euripides' *Bacchae*, one of Ovid's sources, Dionysos has Cadmus and Harmonia 'turn into snakes – / [...] and lead great armies' (1335f.).[38] That Ovid refrains from repeating this rather Shakespearean ending suggests the importance of his twist. Cadmus voluntarily asks for his transformation agreeing to take it as a lesson – a rare case in the *Metamorphoses*. Maybe this is too good an ending to be taken seriously. With respect to the role of the snake in Augustan iconography as a symbol of an eternal time of constant growth – as shown, for instance, on the frieze of the Ara Pacis – Ovid's decision to pacify Cadmus might also contain a grain of irony. The episode leaves its unfinished epilogue on the skin of its readers.

III

Shakespeare's *Macbeth* shares the threefold connection of warfare, agriculture and theatre with Ovid's Cadmus episode. There are even specific allusions to the Cadmus story in the play. As has often been noted, *Macbeth* contains numerous images of growth, from 'the seeds of time' (1.3.58)[39] Banquo mentions when he and Macbeth meet the three sisters, to Duncan's address to Macbeth: 'I have begun to plant thee, and will labour / To make thee full of growing' (1.4.28f.), as well as notions of 'roots', 'seeds', 'barrenness' and 'ripeness' later in the play (e.g. 3.1.5, 60–1, 69; 4.3.240–1, 5.3.41).[40] Macbeth speaks of '[t]he future in the instant' (1.5.58). Ross tells Macduff in act IV that 'your eye in Scotland / would create soldiers' (4.3.187f.). In act I, the Lady urges Macbeth to 'look like the innocent flower, / But be the serpent under't' (1.5.65–66). Two of Macbeth's speeches in act III contain images of a snake and its teeth (3.2.14–16, 3.4.27). In her book on metaphor in Shakespearean drama, Maria Franziska Fahey links Macbeth's wish that the Lady would 'Bring forth men-children only; / For thy undaunted mettle should compose / Nothing but males' (1.7.73–5) to the Cadmus episode, recognizing puns on male/mail, undaunted/ undented, mettle/metal.[41] Moreover, Fahey observes that 'Shakespeare's staging of the armed head emerging out of the Witches' cauldron' recalls Ovid's curtain image.[42]

These allusions provide the theme of birth from a single origin ('none of woman born' 4.1.79) with a mythological foundation, relating it to the notion of autochthony.[43] Like Cadmus, Macbeth kills in surroundings associated with a *locus amoenus*. 'This castle hath a pleasant seat' (1.6.1), Duncan states on the way to Dunsinane; Banquo expands on the fertile and delicate character of Macbeth's estate (1.6.3–10). In both cases, the allegedly peaceful grounds are agitated after the killings. In *Macbeth,* Lennox, the Old Man, and Ross famously speak of the 'feverous' earth (2.3.60–1) and horses about to '[m]ake war with mankind' (2.4.18). Macbeth anticipates his share in those turbulences right after the first encounter with the three sisters calling his possible action to promote his career a 'stir' (1.3.147), the English version of *movere.*

I would like to include one of the play's most pivotal moments into this imagery which conflates warfare, nature and cultivation: the marching of Birnam Woods towards Dunsinane. As I seek to show, the moving forest is not only a strong image for Macbeth's defeat, but has wider implications as a large-scale case of deforestation. The appearance of the trees on stage creates a theatrical moment which allows the audience to experience ambivalent connections between nature, myth and politics.

When the three sisters first speak of the wood 'coming against' Macbeth,[44] Macbeth rejects their prophecy. Yet, at the same time, he makes the image concrete. He transforms the prophecy into the question 'Who can impress the forest, bid the tree / Unfix his earth-bound root?' (4.1.94–5). On the surface, this is a rhetorical question, but it is much more detailed and explicit than the sisters' words. He unwittingly enlivens the image, which will become tangible towards the end of the play. There, 'impress' evolves into its full meaning as a military, an aesthetic and an affective term.[45]

Macbeth is familiar with the idea of animate landscape beyond, and presumably before, his encounters with the three sisters. In act III, after having seen the ghost of Banquo, Macbeth says:

> Stones have been known to move, and trees to speak;
> Augures, and understood relations, have
> By maggot-pies and choughs and rooks brought forth
> The secret'st man of blood.
>
> 3.4.121–4

Like many Shakespearean characters, Macbeth is well versed in the *Metamorphoses*, which is a source for this passage. Elsewhere in the play, elements from Ovid's Medea and

Orpheus resonate in Macbeth's speeches. Macbeth closely follows the words of Medea from Arthur Golding's 1567 translation of the *Metamorphoses* in her famous speech about the powers of nature (and her own power over nature) when he describes the art of the three sisters:[46]

> I conjure you, by that which you profess,
> Howe'er you come to know it, answer me;
> Though you untie the winds and let them fight
> Against the churches, though the yeasty waves
> Confound and swallow navigation up,
> Though bladed corn be lodged and trees blown down,
> Though castles topple on their warders' heads,
> Though palaces and pyramids do slope
> Their heads to their foundations, though the treasure
> Of Nature's germen tumble all together
> Even till destruction sicken, answer me
> To what I ask you.
>
> 4.1.49–60[47]

Medea's speech is also present in Macbeth's above-quoted verse on the apparition of Banquo's ghost: 'Stones have been known to move, and trees to speak' ('both stones and trees do draw', Golding's Medea says). Macbeth's question 'Who can impress the forest?' anticipates Medea's 'Whole woods and forests I remove'. Another of Medea's lines, 'I call up dead men from their graves,' materializes in the ghost of Banquo. There is also a formal parallel between Golding's Medea and Macbeth: the couplet rhyming, so characteristic of Golding's translation (although not unusual in Shakespeare's plays).

In Book 10 of the *Metamorphoses*, having lost Eurydice for good, Orpheus sits on a hill, green, but without shade, playing his lyre. Enchanted by his song, trees of all kinds form a grove around him, providing shade and forming an audience, together with animals and stones. In the opening of Book 11, moments before the maenads attack the singer, Golding subtly connects this orphic idyll of animate landscape to Medea's vehement verses. Orpheus 'draws both stones and trees against their kinds,[48] Golding writes, in almost the same wording he uses in the Medea episode ('both stones and trees do draw'). Medea's tacit presence in the Orpheus episode underlines the temporary and fragile character of Orpheus's lyric theatre, of all theatres. They are and remain connected to the theatre of Dionysian violence the maenads are about to stage with Orpheus's body. Orpheus's idyllic music theatre – a 'utopia of civilization'[49] it has been called – turns into a theatre of war. Golding's 'round about' of 'hungry hounds' in Ovid is a *theatrum*, a dark re-enactment of the peaceful tree audience. The trees, Ovid indicates, stand for stories of suffering and are products of metamorphoses. They do not only provide shade they are also shadows of former existences.[50] The repetitions and metamorphoses continue with the maenads. Bacchus transforms them into trees to avenge Orpheus's death – suggesting justice but also closing the circle for a darker beginning.

The power to master nature through art connects Macbeth both to Medea and Orpheus. Yet Macbeth possesses neither the theatrical powers of a Medea, nor the poetic power of an Orpheus. Rather than setting the scene or writing the plot, he gets drawn into theatrical situations on the 'bloody stage' (2.4.6) that Scotland has become: in his encounters with the Three Sisters, especially in their '*show*' of apparitions (4.1.110, stage dir.), when the Lady instructs him how to 'perform upon' Duncan (1.7.70) and to convincingly arrange the 'pictures' of the dead (2.2.55), when the dagger and Banquo's ghost appear to him, when the trees approach. He is concerned less with poetic form than with his own emotional and physical shape; his couplet rhyming, which closes decisive scenes, is a means of taking on verbal armour rather than a form of poetry.

For a Shakespeare play, *Macbeth* contains surprisingly few references to concrete historical circumstances. An exception is the Porter scene which alludes to the Gunpowder Plot, but also to agriculture: 'Here's a farmer that hanged himself on th'expectation of plenty', the Porter says (2.3.4f.). An audience used to corn prize fluctuations – Shakespeare himself is said to have been involved in speculations – must have understood this phrase also in its literal sense.[51] Other references to farming are strikingly absent in *Macbeth*. The German dramatist Heiner Müller, an avid reader of both Shakespeare and Ovid, was so appalled by this that, for his 1972 translation of the play, he included a peasant plot, accounting for the play's feudal context; somebody must have built Dunsinane, he held.[52]

Forestry, too, is only implicitly present in the play which remains suspiciously silent about the concrete conditions of the woods. However, when Birnam Wood marches toward Dunsinane, the contemporary audience did not only watch a medieval war scene they could translate, as has been suggested, into 'the ritual bringing in of spring'.[53] The impressive presence of the trees must have also reminded them of the fatal relation between warfare and forestry. The forest in Shakespeare's time was a vibrant political, economic and ecological issue. A chapter in Holinshed's *Chronicles* comments on the severe loss of woods in Britain since the Middle Ages due to 'the industrie of man', to agriculture, and hostile destruction in wars.[54] Recently, Vin Nardizzi and Anne Barton have investigated the cultural, economic, political and aesthetic dimensions of the forest in Shakespeare.[55] The exploitation of the British forest began with the Roman conquest in the first century CE but has its peak in Elizabethan times when large amounts of timber were needed to build ships and houses and to provide energy: 'In what many economic historians now regard as the real industrial revolution, furnaces for producing iron (largely used for arms production), copper smelting, glass and salt works all began quite literally to burn up the woods', Barton writes.[56] Nardizzi traces the enormous increase in wood prices which made 'wood products [...] some of the most expensive items a consumer purchased'.[57] The monarchs Elizabeth, James and later also Charles sold off parts of the royal forests to pay for their Irish wars.[58]

Macbeth's servant speaks of 'ten thousand' soldiers approaching Dunsinane (5.2.13; see also 4.3.134 and 191), and Malcolm has 'every soldier hew him down a bough' (5.4.4). The dramatic dialogue suggests a serious scale of deforestation. The play hides these concrete constellations by turning the marching trees into a mythic image. Still, the 'reality effect' that the trees produce impedes solely symbolic readings – even more so in a theatre which mainly creates settings through dialogue – and which itself is made of timber. Nardizzi has emphasized the effects of the wood shortage around 1600 on theatre buildings.[59] Testifying to current events of warfare deforestation,[60] the trees physically draw politics into the theatre and remind of the fact that the political involves concrete bodies not only of kings, queens and soldiers, but also of trees and thus eventually of all life on earth.

Barton emphasizes that Greek *hyle*, and to a certain extent also Latin *silva*, does not only mean 'forest' and 'wood', but also 'Chaos: primordial matter, shapeless, and with only the potential of forms'.[61] With the trees on stage, *Macbeth* ends in an ambivalent image of metamorphosis. The trees may have symbolized renewal and growth, but also delivered a physical impression of shapelessness, even deformation, and thus the opposite of the stable political commonwealth the crowning of Malcolm promises at the end of the play. From this perspective, the closing scenes of *Macbeth* indicate that the play moves beyond a tragedy about greed, pride and corrupt leadership which a new and better leader is about to solve. Another tragedy looms up in these final scenes: one in which nature is neither an obstacle for individual success, nor a mirror for political conduct, but in which it has been forced to become an accomplice of belligerent politics. The soldiers covered in leaves may be read as a mythical image of nature fighting Macbeth back. However, such a reading occludes the fact that this image, too, originates in a violation of nature: deforestation. *Macbeth* thus might not only be concerned with the 'development of an agrarian economy in which nature as a resource must be separated from nature as a representation'.[62] The deployment of Birnam Wood combines both uses of nature; it exploits nature both as a material and a mythical source.

IV

Ovid's Spartoi and Shakespeare's Birnam Wood interlace mythical and historical material in dense constellations, which oscillate between the literal and the figurative and which have multi-layered experiential foundations. The layers continue into the future. In the eighteenth century, Giambattista Vico read the Cadmus myth straightforwardly as an allegory of archaic agrarian politics, even of an archaic class struggle based on agriculture. His allegoresis, which is part of the section 'Epitomes of poetic history' in *The New Science*, reads as follows:

> He kills the great serpent; he deforests the great ancient forest of the Earth. He sows the teeth of this serpent; this is a fine metaphor [...] for the hard curved wood which prior to the discovery of the use of iron must have served as the teeth of the earliest ploughs, and with these, which are still called "teeth," they ploughed the earliest fields of the world. He casts a great stone – that is, the hard earth that the clients, or familial servants, wished to plough for themselves [...]. Born from the furrows were armed men; these represent the heroic contests over the first agrarian law in which [...] the heroes emerged from their grounds – which is to say that they were the lords of those grounds – and united in arms against the plebeians. And they fought not among themselves but with the clients rebelling

> against them. And the furrows signify those orders in which they united and by which they formed and settled the first cities on the basis of arms [...]. And Cadmus changes into a serpent; thus does the authority of aristocratic senates come into being, or, as the most ancient peoples of Latium would have said, *Cadmus fundus factus est* ('Cadmus has become the land'), and, as the people of Greece said, Cadmus changed into a Draco, who wrote laws in blood. [...] [T]he myth of Cadmus contains many centuries of poetic history.[63]

Vico claims that early Egyptian and Greek history was marked by a struggle between the 'heroes', i.e. the ruling aristocrats, and subdued and exploited 'servants', the plebeians of archaic times.[64] Myth, Vico holds, is part of the symbolic language of aristocrats who represent themselves as heroes in the myths.[65] From this perspective, the Cadmus myth becomes an aristocratic celebration of the submission of the plebeians. Vico might also count Roman Republican and early Imperial times among the 'many centuries of poetic history' the episode speaks about. The class battles he reads into the myth also reflect the battles around *ager publicus*, which often was hardly public, but available only to the 'great capitalists' as Max Weber calls Roman landowners in his famous study on Roman agrarian history.[66] Moreover, Vico writes *The New Science* in the first half of the eighteenth century when Europe starts experiencing transformations in agriculture and class structures due to the gradual advent of industrialization.[67] Contemporary agrarian politics must have been of interest to Vico: recent research in interrelations between Enlightenment politics and agriculture has shown that Vico's *New Science* had been a source for the Physiocrats, a group of agrarian philosophers in the circle of François Quesnay and Anne Robert Jacques Turgot.[68] While the kind of concrete historical connections that Vico locates in the Cadmus myth remain latent in Ovid's telling of the story, an alarming feeling of historical continuity concerning the relationship between agriculture, political equality and economic distribution comes to the fore through these possible historical contextualizations. That class issues remain crucial for the discussion of agricultural aspects of the anthropocene, adds to this feeling.[69]

Heiner Müller, whose materialist adaptation of *Macbeth* I mentioned above, kept reworking instances of what he called a 'war of the landscapes' for the twentieth century. His plays strengthen the political and ecological layers of Ovidian and Shakespearean imagery towards a critique of imperial exploitation and address Vico's class issues from a global perspective. In his 1979 play *Der Auftrag* (The Task/The Mission) about a failed revolutionary upheaval in late-eighteenth-century Jamaica, Müller has the former slave Sasportas speak the following lines:

> Der Aufstand der Toten wird der Krieg der Landschaften sein, unsre Waffen die Wälder, die Berge, die Meere, die Wüsten der Welt. Ich werde Wald sein, Berg, Meer, Wüste. Ich, das ist Afrika. Ich, das ist Asien. Die beiden Amerika bin ich.
>
> *The upheaval of the dead will be the war of landscapes, our weapons the forests, the mountains, the seas, the deserts of the world. I will become forest, mountain, sea, desert. Me, this is Africa. Me, this is Asia. I am the two Americas.*[70]

Sasportas' lines create a new myth of the oppressed metamorphing into landscape to organize their upheaval on a new scale. The ecopolitical dimension of the sentences, which were less urgently present to Müller in the late 1970s, is gradually developing its potential as Müller's plays become canonical.

'We are all involved in this disaster' (1304) Euripides has his Cadmus say towards the end of the *Bacchae*, before he becomes a snake and, along with Harmonia, is to 'lead great armies' (1336). With their afterlives in mind, Ovid's Cadmus story and Shakespeare's *Macbeth* remain starting points for approaching the scandal of repetition in the shared history of warfare and agriculture. The (meta-)theatrical elements in the texts suggest that we need a common ground to experience our involvement in this scandal not only intellectually, but also on visceral levels. As a provisional, tentative space in which bodies come together in collective presence and experiment with various forms of distance and involvement, theatre is an especially apposite platform for this experiential practice. Two millennia after Ovid's Cadmus and four centuries after *Macbeth*, we are beginning to recognize the scandal of repetition as a scandal of belatedness. In his manifesto on ecopolitics, *Down to Earth*, Bruno Latour imagines Birnam Wood moving off to seek vengeance: 'We have benefited from every resource; now these resources, having become actors in their own right, have set out, like Birnam Wood, to recover what belongs to them.'[71] Latour suggests that the forest in Shakespeare's play comes against Macbeth to revenge his breach of nature. The discourse of deforestation that haunts *Macbeth* indicates that this might be a romantic idea. Nature does not 'fight back'; it is showing its wounds in the global theatre of experience that we still hold at a distance by viewing it as if on a painted curtain.

Notes

1. This concept of experience between pathos and response was developed by Waldenfels 2002, drawing on Merleau-Ponty.
2. Cf. Blumenberg 1979, esp. 59.
3. On the experiential quality of the arts – that is, the material, somatic and affective foundations of artworks and the ways in which their interplay, together with the recipients' somatic awareness, evoke forms of aesthetic experience, cf. Shusterman 2018; Waldenfels 2010; and Formis 2009.
4. Fuchs 2017 has investigated collective learning on experiential grounds and has shown that body memory records feelings and 'techniques of the body' (Marcel Mauss), and thus interlaces individual experiences and (even long-lasting) cultural codes.
5. As a sample of this vast and growing field of research, see Wreford, Moran and Adger 2010; and Ruddiman 2003, who argues that the Anthropocene began with the invention of agriculture.
6. Hardie 2002a: 3 has described how Ovid creates such effects of a tangible presence through a poetics of 'duplicity [...] which equivocates between absence and presence and which delights in conjuring up illusions of presence'. How deeply Shakespeare's theatre is indebted to Ovid's style and themes (from love to myth and politics) has received increasing attention in recent decades. Among the more comprehensive studies are: Bate 1993; Enterline 2000; Taylor 2000; Starks-Estes 2014; and Reid 2018.

7. On the relationship between plasticity and intense aesthetic experience, see Largier 2013.
8. Hanson 1998: 11.
9. Hanson 1998: 13.
10. Hanson 1998: 17. Cf. also Thorne 2007.
11. Goldsworthy 2000: 28.
12. Maschek 2018: 145.
13. Maschek 2018: 157ff.
14. Cf. Weeda 2015: 59–84.
15. Maschek 2018: 173.
16. Maschek 2018: 172ff.
17. English translation by Johnson 2009.
18. Melville's English translations are taken from Ovid 2009: 51.
19. Cf. Keith 2017: 243–7.
20. Hardie 1990: 225.
21. Harrison 1992: 55.
22. Hardie 1990: 227.
23. Bömer 1969, 478 ff.
24. The connection between Ovid's curtain image and this passage in Virgil has become topical in Ovidian scholarship. Cf. e.g. Giusti 2019: 113ff.; Schmitzer 1990: 142; Bömer 1969: 478ff. Hardie 1990: 230 mentions a different Virgilian source for Ovid's image, the 'simile of the sea rising gradually before the blast of storm-winds' in the Aeneid (7.528–30).
25. Dench 2005: 205ff.; Radke-Stegh 1978: 68ff.
26. Cf. Welch 2007: 186 and also 27ff.
27. Welch 2007: 77, 119, 187; Maschek 2018: 195, 202.
28. Maschek 2018: 195; cf. also Welch 2007: 80ff., 102–8. On seating rules as means of social stability cf. also Rehm 2007: 197ff.
29. Welch 2007: 9, and 165 ff., where she notes that, from the first century CE, Greek theatres (including the Theatre of Dionysus in Athens), were converted into venues for gladiatorial games.
30. Rehm 2007: 193ff.
31. Cf. Welch 2007: 58–71. With respect to mythological and historical layering, one could cite the Forum Boarium, Rome's cattle market, where gladiatorial fights took place and where Hercules was said to have slain Cacus.
32. Radke-Stegh 1978: 69.
33. Feldherr 2010: 160–98 describes several similar moments in Book 3 of the *Metamorphoses*. Cf. esp. p. 181ff. on the tension of involvement and distance in the curtain passage. Hardie 2002a: 167–70 also comments on the importance of theatricality, especially a dionysian theatricality in *Met.* 3. For further examples of theatricality in Ovid, cf. Hardie 2002b: 38–42; and Curley 2013.
34. Feldherr 2010: 168 has emphasized 'the reciprocity between Ovidian visualization and contemporary Roman spectacle'. In this context, see also Feldherr 1997 (esp. p.44) for a reading of amphitheatricality in the Actaeon episode.
35. Cf. Keith 2002b: 262ff.; for tragic elements in the episode (channelling tragic elements in Virgil's *Aeneid*), see Hardie 1990: 226. For Thebes as a *topos* of Greek drama, see Zeitlin 1990.

36. Cf. Flaig 2007: 85.
37. Feldherr 1997: 29 has shown that looking/watching and being watched are densely intertwined in Book 3; he describes several instances of a 'reversal in visual roles'. See also Feldherr 2010: 180.
38. Translation of Euripides, *Bacchae* taken from Walton 1988: 145.
39. All quotations from *Macbeth* taken from the edition by Clark and Mason 2015.
40. Gowers 2011: 101ff. comments on these and other quotes from *Macbeth* with respect to the relation of tree imagery to ideas of dynasty and genealogy. Scott 2014: 121–49 discusses images of barrenness and cultivation in *Macbeth* as attempts to use nature for self-fashioning.
41. Fahey 2011: 89 ff. Brooke 1990: 121 had already noted these puns.
42. Fahey 2011: 90.
43. Fahey 2011: 91.
44. 'Be lion-mettled, proud, and take no care / Who chafes, who frets, or where conspirers are. / Macbeth shall never vanquished be, until / Great Birnam wood to high Dunsinane Hill / Shall come against him' (4.1.89–93).
45. On the verb's semantic ambivalence see also Harrison 1992: 103.
46. For Golding's translation of Medea's speech, see Forey 2002: 208ff.
47. It has often been observed that other parts of Ovid's Medea episode provide ingredients for the witches' cauldron, and that the Ovidian and the Senecan Medeas may have influenced the character of Lady Macbeth. Macbeth's relation to Medea has received little critical attention.
48. Forey 2002: 321 for Golding's translation of *Met.* 11.1–2.
49. Döring 1996: 39 (my own translation).
50. Döring 1996: 45. See also the subtitle of Harrison 1992.
51. On the topic of grain in Shakespeare, see Archer, Thomas and Turley 2015.
52. Müller 2001 (1972): 573 on the peasants who built the castle.
53. Barton 2017: 53.
54. Holinshed 1587: 212.
55. Nardizzi 2013; Barton 2017.
56. Barton 2017: 6.
57. Nardizzi 2013: 10.
58. Barton 2017: 7.
59. Nardizzi 2013: 15–20. Throughout this book, Nardizzi founds his readings of Elizabethan plays on the wooden materiality of the playhouses.
60. Nardizzi 2013: 28–31, 59–83 and 112–35 elaborates on such moments in Shakespeare's *The Tempest* and *The Merry Wives of Windsor*, but mentions the trees in *Macbeth* only cursorily.
61. Barton 2017: 49. This semantic dimension reflects in the image of the forest as a place of dis-order, shapeshifting and transformation (on which, see Barton 2017: 53).
62. Scott 2014: 149.
63. Vico 2020: 287ff. (section 679).
64. Cf. Vico 2020: 17 and 59 (sections 20 and 81). See also Vico 2020: 238 (section 583) for his account of Roman agrarian rebellion by the plebeians: 'They must have been the Ixion who forever turns a wheel, and the Sisyphus who pushes upward the rock which Cadmus threw,

the hard earth which returns once it reaches the top – this is retained in the Latin expressions, vertere terram ['turning earth'], for cultivating the land, and saxum volvere ['turn the stone'], for arduously performing a long and difficult task. It was on account of all this that the familial servants must have rebelled against the heroes.

65. Cf. Vico 2020: 9 and 49 (sections 7 and 52).

66. Cf. Weber 1891: 129. Weber describes a system of exploitation between the landowners and dependent landworkers and peasants at Weber 1891: 233ff.

67. While there were few changes in Italy at that time, England, for instance, experienced the enclosure movement which significantly diminished common land.

68. Cf. Richter 2015: 310–24.

69. Cf. Godfrey and Torres 2016: 141–200.

70. Müller 2002 (1979): 38–40 (my own translation). Similar imagery recurs in other texts by Müller that touch on the *Metamorphoses* which, alongside Shakespeare's reading of Ovid, are pivotal to Müller's poetics. Cf. Müller 2005: 334–7.

71. Latour 2018: 103ff.

EPILOGUE: A GLOBALLY WARMED *METAMORPHOSES*

John Shoptaw

In my poetic sequence 'Whoa!' I rendered Ovid's tale of Phaethon as an allegory of global warming. Early in his *Metamorphoses* (Books 1 and 2), Ovid tells how the illegitimate son of the Sun is granted a wish by his father and foolishly demands to drive his Chariot for a day – a joyride with disastrous consequences for the young rider and for the Earth. It is a tale that, read in our Anthropocene epoch, already has allegorical potential. But the more time I spent with it, the more I was taken by what the whole story of Phaethon's ride would enable me to do. In *The Great Derangement: Climate Change and the Unthinkable*,[1] the novelist Amitav Ghosh provocatively declares that climate change cannot be represented by mainstream fiction (or presumably, narrative poetry) because it lies beyond the bounds of narrative realism. Where it does get represented, in the sub-genre of science fiction known as speculative or climate fiction, it tends to happen elsewhere – on another planet, in another time, or in a parallel universe. (In his 2020 novel, *The Ministry for the Future,* Kim Stanley Robinson evades this criticism by setting events in our own planet's near future.) Phaethon's global chariot ride, fantastical though it certainly is, afforded me a view of global warming as it is already happening – the deranged globe of my poem is literally our contemporary earth. For my 'day' I chose the period around the turning of the year from 2016 to 2017, when I began thinking about the poem, and 'Whoa!' includes nothing that didn't actually occur during that time. (My title comes from Phaethon's panicked address to the horses, whose names he has forgotten, but it also expresses our everyday awe before the sublimely abnormal or unreal. And it may also echo 'woe', which some readers have heard, though I hadn't at the time.)

One challenge to rendering global warming as a whole, over and above its extreme weather events, is its heterogeneity. When treating overwhelming events or conditions, writers tend to resort to synecdoche, giving the reader an experience of the whole through a representative locale or group: a town, a wood, an island, a family, a band of (human or nonhuman) survivors, an outcast seer and truth-teller (cf. the film *Don't Look Up*), and so on. Compare global warming with the Covid-19 pandemic. Where this plague exhibits much the same symptoms everywhere, global warming manifests itself differently over the planet: sunny-day flooding, violent hurricanes and tornados, droughts, heat waves, rain bombs, polar vortex southern blizzards, glacial melts, the migration of native and the invasion of non-native species, human resettlement etc. My own California Bay Area has been invaded by tree-bark beetles, which have killed over a hundred million California trees. These dead trees, exacerbated by a long-term drought, along with uncleared brush, have fuelled violent forest fires and widespread toxic smoke.

(We strapped on N-95 masks before Covid-19 arrived.) Since no one place can be representative, the synecdochic strategy can only produce stories that, while moving and important, may well seem exotic to readers experiencing another array of extreme weathers.

While I cannot, of course, depict the entire world (all literature is selective, and hence synecdochic), Phaethon's scorching daytrip did give me a way of representing a number of the far-flung effects (along with some causes) of our planetary climate crisis. If the Sun had been driving his own chariot badly, I would have had no Ovidian poem. By handing the reins to Phaethon, Ovid gave me a human view on our stressed planet that scientific graphs and satellite photos lack. I patterned my Phaethon, whom I called Ray, after Donald Trump, the impetuous heir of Sol. (Trump's 2017 announcement that he would pull the US out of the 2015 Paris Climate Accords helped spur 'Whoa!' into being.) My poem begins with the frustrated son of the Sun pacing his unnamed 'hotel corridors' till he comes across 'his old wall map thumbtacked / over some double doors'. This doorway, ornamented with a Mercator-projected *Boy's Life* world map, is my transformation of the entry to Sol's palace, with its silver double doors ('bifores . . . valvae') upon which Vulcan has ecphrastically wrought the flattened globe. Folding up the wall map and entering, Ray finds himself before a sumptuously ornamented escalator (reminiscent of Trump Tower's golden escalator, which Trump and his family descended, as if from Olympus, to launch his 2016 presidential campaign). Ray takes the escalator up to what he imagines would be 'the Sun's penthouse' (For 'Whoa!' I transformed Ovid's flexible hexameters into emphatic four-beat lines.)

> With a little lurch the escalator began
> its quickening glide. Now this was first-class.
> No huffing required. Simply the assumption
> of an attitude.

Almost inevitably, Trump's predecessor, the climate-conscious Barack Obama, got cast as Sol, so I turned the Sun's palace or penthouse into an all-season Marvellian garden tended by the 'ambling photosynthesizer [who] passed / the fragrant hours of the zodiac'. Perfect as Trump was as a model for my charioteer – foolhardy, blissfully ignorant, posturing and, in the event, cowardly – my project was hampered by his oblivious self-regard: it is hard to imagine him looking with much interest at the world suffering below him. So I retro-fitted Ray's solar chariot, still flown by four pegasi, with his interactive *Boy's Life* map and a pair of telescopic goggles, so his view might be enhanced to rival that of the encyclopedic Ovid.

My poem emulates Ovid's ambition by widening his survey to match our own calamitous global view. After his untoward climb into the constellations (my climate satellites) and a plunge below the moon, Ray finds himself above the cliff-side woods of a forest-fired California, and a consequent mudslide, 'upon which Hope too gave way'. I then cross the Pacific, veering north from Australia, 'where coal bulkers / rocked in port', to India with its drying Ganges, China with its 'chemical clouds', and Russia, whose LNG (liquid natural gas) pipe lines I figured as a 'pipe-organ' branching into the Middle East and Europe. From here I proceeded not only westward but downward by tracing my rivers and lakes to their glacial sources:

> . . . the Rhône glacier, blanketed in white
> reflective blankets by Swiss teams
> reduced to hospice nurses, its darling
> Lake Geneva's temperature rising

I closed this part with a catalogue of melting glaciers ('Pasterze Hornkees Hintereis Ferner'), which I followed, a few parts later, with a catalogue of coal-powered plants. Yet as I proceeded with my transformations, I was nagged by a crucial difference between Ovid's tale and my rendition: Ovid's disasters were made up while mine really happened (and mostly still are). Like any translator, I had the benefit of hindsight; but Ovid had the advantage of foresight. He knew before he began that his tale of a metamorphic global doomsday would be followed by another tale and another, just as the charred Arcadia would be regreened and reanimated by Jupiter (till Callisto distracts him). If 'All things change', as his Pythagoras claims, change itself, like Ovid's poem, is perpetual. His terrified and hapless Phaethon, then, is for Ovid just another character, a callow youth who is occasionally ridiculous and only in his epitaph's telling sublime. I, on the other hand, don't know how global warming will come out. And while I selected and arranged catastrophes to align with Ovid's, I couldn't change or reverse them. Further, while I similarly use Ray for much needed satire and even comic relief, I couldn't simply soar above him in ironic detachment. His fear and confusion were too close for comfort. I confess my relative incapacity in the seventh part of 'Whoa!' where I begin with a desperate wish:

> What I'd give for a swig from the Hippocrene
> hiding somewhere up on Mount Helicon
> where Pegasus sank his hoof and the water
> bucked, or even a swallow of Arrowhead
> from the San Bernardino Mountains bottled
> in clear recyclable petroleum plastic.
> The *Metamorphoses* looks down on its clueless charioteer
> while its mountains and rivers unroll in perpetual
> hexameters serene as a Chinese handscroll.
> I'm not up to it. I have no command over
> my recalcitrant materials, so far beyond my
> power to remedy or return to their courses
> that despite my getting a feel for my rendition
> I know I'm also a petrified rider
> whose classic vehicle is running out of fluids.
> It's all I can do to follow the circumnavigation

to its downfall in my stamping four-horse verses
on a long day when it's always noon.

Rereading these lines, a couple of things come back to me. I was pleasantly surprised when writing to learn that 'recalcitrant' sprang from *buck*, and I barely refrained at the end from alluding to one of my favorite films, *High Noon*.

Disgusted with Europe's solar panels and wind turbines ('Those eyeless sunshades, / those propellers propelling nothing nowhere'), Ray's chariot quits Europe (with its Climate Accords) and veers south into Africa's Sahel, threatened by the advancing Sahara ('Sahel' in Arabic means *coast*). Then the 'horsebirds' lurch back north across the Atlantic, with its wobbly ocean currents, toward a freakishly snow-covered Manhattan, above which Ray scoffs, 'Where's the heat?' (echoing Walter Mondale's 'Where's the beef?'). Ovid foreshadows the end of Phaethon's ruinous ride with an appeal to Jupiter by Tellus, Mother Earth. Her appeal was what first drew me to Ovid's disturbing, epic-scaled tale since it resonates so profoundly with the distress and outrage many of us feel today. For Tellus, the assaults on her own body are at once global and personal, comprehensive and felt. I needed her emotional intensity, which is something Phaethon's chariot ride largely lacks. Through her climactic, operatic appeal in the poem's eleventh part, I was able to embody some of the painful and wrenching metamorphoses which ravaged our own planet:

My Kentucky coal-hills are lobotomized,
and coal-ash ponds are blinding my lakes
and coating my cold freshwater tongues.
My seas are bowls of plastic soup

Moving from mountain to stream, to the oceans, to the north and south poles, my Ovidian Tellus can articulate the breadth and force of her outrage, agony and distress, both present and looming. Since in my view of the world there's no divinity to answer her prayer, I leave it to Mother Earth herself (that is, to us responsible humans) to deal with the heedless charioteer (and his 'fossil lordlings'). Ovid's deus ex machina may seem unbelievable, but I think we can and need to entertain its possibility. After all, we did just unseat Ray's climate-denying, pollution-promoting model.

Ovid's thunder-bolted pretender crashes down into Italy's river Po, where his body is tended and elegized by Hesperian (Italian) Naiads. In 'Whoa!', with admittedly more than a little poetic justice, Ray comes sailing down into a South Dakota reservoir on the Missouri River, under which snakes a tar-sands oil pipeline, where he is memorialized by a Sioux poet:

A Standing Rock Sioux nymph
spray-painted his epitaph on the concrete dam:

Here's where the son sank.
Far did he fall for so deep a fool.

> Then she rejoined her sisters downstream,
> and on the legendary gravesite of Sitting Bull
> they stood up a row of sticks at the shoreline.
> The twigs dug down and branched right up
> into a midwinter flurry of cottonwood seed-fluff.

A nearer translation of the epitaph clarifies my changes:

> Hic situs est Phaethon currus auriga paterni
> Quem si non tenuit magnis tamen excidit ausis.

> *Here lies Phaethon, charioteer of his father's chariot;*
> *if he didn't hold onto it, he failed having greatly dared.*

Ovid's epitaphists are more admiring than my poet, whose admiration, if it exists, is bitter. While I was composing my poem, members of the Standing-Rock Sioux tribe did indeed plant native cottonwood trees nearby on Missouri river banks (I no longer find the online posting). So too, transfiguring Phaethon's tree-changed, amber-teared sisters, I ended 'Whoa!' by transplanting along the dammed but still flowing Missouri these quickly growing, festively seeding riverside trees.

Note

1. Ghosh 2016.

BIBLIOGRAPHY

Abram, D. (1997), *The Spell of the Sensuous: Perception and Language in a More-Than-Human World*, New York, NY: Vintage Books.

Adluri, V., and J. Bagchee (2012), 'From Poetic Immortality to Salvation: Ruru and Orpheus in Indic and Greek Myth,' *History of Religions* 51.3: 239–61.

Aït-Touati, F., and E. Coccia, eds (2021), *Le Cri de Gaïa: Penser la terre avec Bruno Latour*, Paris: Éditions La Découverte.

Alfonsi, L. (1958), 'L'inquadramento filosofico delle *Metamorfosi*' in L. G. Herescu, ed., *Ovidiana*, Paris: Les Belles Lettres, pp. 265–72.

Allen, D. (1964), 'Orpheus and Orfeo: The Dead and the Taken,' *Medium Ævum* 33.2: 102–11.

Alvar, J. (2008), *Romanising Oriental Gods: Myth, Salvation, and Ethics in the Cults of Cybele, Isis and Mithras*, ed. and trans. Richard Gordon, Leiden: Brill.

Angremy, A. (1983), *La Mappemonde de Pierre de Beauvais*, *Romania* 104.415: 316–50 and 104.416: 457–98.

Apuleius (1915), *The Golden Ass*, trans. William Adlington, spelling modernized, London and New York: Heinemann and Macmillan (Loeb Classics Library).

Archer, J. E., H. Thomas and R. Marggraf Turley, eds (2015), 'Reading Shakespeare with the grain: Sustainability and the hunger business,' *Green Letters: Studies in Ecocriticism* 19.1: 8–20.

Aristophanes (2000), *Birds. Lysistrata. Women at Thesmophoria*, ed. and trans. Jeffrey Henderson, Cambridge Mass.: Harvard University Press (Loeb Classics Library).

Armstrong, A., and S. Kay (2011), *Knowing Poetry: Verse in Medieval France from the Rose to the Rhétoriqueurs*, Ithaca and London: Cornell University Press.

Armstrong, R. (2006), *Cretan Women: Pasiphae, Ariadne, and Phaedra in Latin Poetry*, Oxford: Oxford University Press.

Armstrong, R. (2019), *Virgil's Green Thoughts: Plants, Humans, and the Divine*, Oxford: Oxford University Press.

Asmis, E. (1982), 'Lucretius' Venus and Stoic Zeus,' *Hermes* 110.4: 458–70.

Baeten, S. (2020), *Birds, Birds, Birds: A Comparative Study of Medieval Persian and English Poetry*, Munich: Utzverlag.

Baker, C. (2003), 'De la paternité de la version longue du *Bestiaire*, attribuée à Pierre de Beauvais,' in B. Van den Abeele, ed., *Bestiaires médiévaux: Nouvelles perspectives sur les manuscrits et les traditions textuelles. Communications présentées au XVe Colloque de la Société Internationale Renardienne (Louvain-la-Neuve, 19–22.8.2003)*, Louvain-la-Neuve: Publications de l'Institut d'Études Médiévales, Université Catholique de Louvain, pp. 1–29.

Baker, C., ed. (2010), *Le Bestiaire: Version longue attribuée à Pierre de Beauvais*, Paris: Champion.

Baldwin, William (1547), *A Treatise of Morall Phylosophie, Contayning the Sayings of the Wyse. Gathered and Englyshed by Wylm Baldwyn*, England: Whitchurche.

Barchiesi, A. (1994), *Il poeta e il principe: Ovidio e il discorso augusteo*, Bari: Laterza. English translation (1997), *The Poet and the Prince: Ovid and Augustan discourse*, Berkeley: University of California Press.

Barchiesi, A. (2001), 'The Crossing,' in S. Harrison, ed., *Texts, Ideas, and the Classics: Scholarship, Theory, and Classical Literature*, Oxford: Oxford University Press, pp. 142–63.

Barchiesi, A. (2002), 'Narrative Technique and Narratology in the Metamorphoses,' in P. R. Hardie, ed., *The Cambridge Companion to Ovid*, Cambridge: Cambridge University Press, pp. 180–99.

Barchiesi, A., ed. (2005), *Ovidio, Metamorfosi (libri I–II)*, vol. I, Milan: Mondadori.
Barchiesi, A., J. Rüpke and S. Stephens, eds (2004), *Rituals in Ink. A Conference on Religion and Literary Production in Ancient Rome held at Stanford University in February 2002* (Potsdamer Altertumswissenschaftliche Beiträge 10), Stuttgart: Franz Steiner Verlag.
Barkan, L. (1986), *The Gods Made Flesh: Metamorphosis and the Pursuit of Paganism*, New Haven: Yale University Press.
Barolsky, P. (2003a), 'Ovid's Colors', *Arion* 10.3: 51–6.
Barthes, R. (1967), 'The Death of the Author', *Aspen* 5+6.
Barton, A. (2017), *The Shakespearean Forest*, Cambridge: Cambridge University Press.
Barton, C. (1989), 'The Scandal of the Arena', *Representations* 27: 1–36.
Bate, J. (1993), *Shakespeare and Ovid*, Oxford: Clarendon Press.
Beasley, M. (2012), *Seriously Playful: Philosophy in the Myths of Ovid's* Metamorphoses, A thesis presented for the degree of Doctor of Philosophy of Classics of The University of Western Australia.
Beaujour, M. (1980), *Miroirs d'Encre*, Paris: Le Seuil.
Bennett, J. (2004), 'The Force of Things: Steps toward an Ecology of Matter', *Political Theory* 32.3: 347–72.
Bergren, A. (1989), '"The Homeric Hymn to Aphrodite": Tradition and Rhetoric, Praise and Blame', *Classical Antiquity* 8: 1–41.
Bernard, S. (2018), *Building Mid-Republican Rome: Labor, Architecture, and the Urban Economy.* Oxford: Oxford University Press.
Bernhardy, G. (1850), *Grundriss der römischen Litteratur*, 2nd ed., Halle: C. A. Schwetschke und Sohn.
Bevis, J. (2010), *Aaaaw to Zzzzzd: The World of Birds, North America, Britain, and Northern Europe*, Cambridge, Mass.: MIT Press.
Bianchi, E., S. Brill and B. Holmes, eds (2019), *Antiquities Beyond Humanism*, Oxford: Oxford University Press.
Blumenberg, H. (1985), *Work on Myth,* trans. Robert M. Wallace, Cambridge, Mass./London: MIT Press.
Bömer, F. (1969), *P. Ovidius Naso, Metamorphosen – Kommentar* (Vol. I, book I – III), Heidelberg: Carl Winter.
Borlik, T. (2016), 'Unheard Harmonies: *The Merchant of Venice* and the Lost Play of *Pythagoras*', *Medieval and Renaissance Drama in England* 29: 191–221.
Breed, B. (2000), 'Silenus and the *Imago Vocis* in *Eclogue 6*', *Harvard Studies in Classical Philology* 100: 327–39.
Bricault, L. (2018), 'Traveling Gods: The Cults of Isis in the Roman Empire,' in J. Spier et al., eds, *Beyond the Nile: Egypt and the Classical World*, Los Angeles: Getty Museum, pp. 226–31.
Bricault, L. (2020), *Isis Pelagia: Images, Names, and Cults of a Goddess of the Seas*, Leiden: Brill.
Brooke, N., ed. (1990), *William Shakespeare: The Tragedy of Macbeth*, Oxford: Oxford University Press.
Burkert, W. (1983), *Homo Necans: The Anthropology of Ancient Greek Sacrificial Ritual and Myth,* trans. Peter Bing, Berkeley: University of California Press.
Burkert, W. (1985), *Greek Religion*, trans. John Raffan, Oxford: Blackwell.
Burkert, W. (2004), *Babylon, Memphis, Persepolis: Eastern Contexts of Greek Culture*, Cambridge, Mass.: Harvard University Press.
Burrus, V. (2019), *Ancient Christian Ecopoetics: Cosmologies, Saints, Things*, Baltimore: Johns Hopkins University Press.
Butler, S. (2015), *The Ancient Phonograph*, New York: Zone Books.
Butler, S. (2019a), 'Is the Voice a Myth? A Rereading of Ovid', in Martha Feldman and Judith T. Zeitlin, eds, *The Voice as Something More: Essays toward Materiality*, Chicago: University of Chicago Press, pp. 171–87.

Butler, S. (2019b), 'What Was the Voice?', in Nina Sun Eidsheim and Katherine Meizel, eds, *The Oxford Handbook of Voice Studies*, New York: Oxford University Press, pp. 3–17.
Bynum, C. W. (2011), *Christian Materiality: An Essay on Religion in Late Medieval Europe*, New York: Zone.
Campbell, B. (2012), *Rivers and the Power of Ancient Rome*, Chapel Hill, NC: The University of North Carolina Press.
Campbell, E. (2020), 'Sound and Vision: Bruno Latour and the Languages of Philippe de Thaon's *Bestiaire*', *Romanic Review* 111.1: 128–50.
Campbell, G. (2003), *Lucretius on Creation and Evolution: A Commentary on the Rerum Natura Book Five, lines 772–1104*, Oxford: Oxford University Press.
Carruth, A. (2013), *Global Appetites: American Power and the Literature of Food*, Cambridge: Cambridge University Press.
Casali, S. (2006), 'The Art of Making Oneself Hated: Rethinking (anti-) Augustanism in Ovid's *Ars Amatoria*,' in R. K. Gibson, S. Green and A. Sharrock, eds, *The Art of Love: Bimillennial Essays on Ovid's* Ars Amatoria *and* Remedia Amoris, Oxford: Oxford University Press, pp. 216–34.
Casali, S. (2007), 'Correcting Aeneas's Voyage: Ovid's Commentary on *Aeneid* 3', *Transactions of the American Philological Association* 137.1: 181–210.
Casanova-Robin, H. (2009), 'Dendrophories d'Ovide à Pontano: la nécessité de l'hypotypose', in H. Casanova-Robin, ed., *Ovide. Figures de l'hybride. Illustrations littéraires et figurées de l'esthétique ovidienne à travers les âges*, Paris: Champion, pp. 103–24.
Cavarero, A. (2005), *For More than One Voice: Toward a Philosophy of Vocal Expression*, trans. Paul A. Kottman, Palo Alto: Stanford University Press.
Centamore, F. (1997), '"*Omnia mutantur, nihil interit*": il pitagorismo delle *Metamorfosi* nell'idea di natura di Bruno,' *Bruniana & Campanelliana* 3.2: 231–43.
Chakrabarty, D. (2009), 'The Climate of History: Four Theses', *Critical Inquiry* 35.2: 197–222.
Chambers, A. B. (1961), '"I Was But An Inverted Tree": Notes Toward the History of an Idea', *Studies in the Renaissance* 8: 291–9.
Chance, J. (2013), 'Re-Membering Herself: Christine de Pizan's Refiguration of Isis as Io', *Modern Philology* 11.2: 133–57.
Chesi, G., and F. Spiegel, eds (2020), *Classical Literature and Posthumanism*, London and New York: Bloomsbury.
Chess, S. (2015), '"Or Whatever You Be": Crossdressing, Sex, and Gender Labour in John Lyly's *Gallathea*.' *Renaissance and Reformation / Renaissance Et Réforme*, 38.4: 145–66.
Chien, J.-P. (2006), 'Of Animals and Men: A Study of *Umwelt* in Uexküll, Cassirer, and Heidegger', *Concentric: Literary and Cultural Studies* 32.1: 57–79.
Ciabaton, M. A. (2020), 'Multiplicity (Intertextual et al.) in *Metamorphoses* 10.560–707: Atalanta's Duplicity Unveiled', *Classical Philology* 115.2: 186–208.
Claassen, J.-M. (2016), 'Seizing the Zeitgeist: Ovid in Exile and Augustan Political Discourse', *Acta Classica* 59: 52–79.
Clark, S., and P. Mason, eds (2015), *William Shakespeare: Macbeth*, London et al.: Bloomsbury.
Clay, D. (2014), 'The Metamorphoses of Ovid in Dante's *Divine Comedy*', in J. Millar and C. Newlands, eds, *Handbook to the Reception of Ovid*, Oxford: Wiley-Blackwell, pp. 174–86.
Clément, G. (2004), *Manifeste du tiers paysage*, Paris: Éditions Sujet/Objet.
Clier-Colombani, F. (2015), 'Les différents programmes iconographiques: Filiations entre les manuscrits de l'*Ovide moralisé*', *Cahiers de recherches médiévales et humanistes* 30: 21–48.
Clier-Colombani, F. (2017), *Images et imaginaire dans l'Ovide moralisé*, Paris: Champion.
Clifford, J. (1986), 'Introduction: Partial Truths', in J. Clifford and G. Marcus, eds, *Writing Culture: The Poetics and Politics of Ethnography*, Berkeley: University of California Press, pp. 1–26.
Coccia, E. (2020), *Métamorphoses*, Paris: Bibliothèque Rivages.
Cole, T. (2008), *Ovidius Mythistoricus: Legendary Time in the Metamorphoses*, Frankfurt: Peter Lang.

Congourdeau, M.-H. (2007), 'L'embryon et son âme dans les sources grecques (VIe siècle av. J.-C.- Ve siècle apr. J.-C.)', *Centre de recherche d'histoire et civilisation de Byzance, Monographies 26*, Paris: Association des amis du Centre d'histoire et civilisation de Byzance.
Connor, S. (2014), *Beyond Words: Sobs, Hums, Stutters and Other Vocalizations*, London: Reaktion.
Coo, L. (2007), 'Polydorus and the Georgics: Virgil *Aeneid* 3.13–68', *Materiali e Discussioni per l'annalisi dei Testi Classici* 59: 192–9.
Cordovana, O. D., and G. Chiai (2017), 'Introduction: The Griffin and the Hunting', in O. D. Cordovana and G. Chiai, eds, *Pollution and the Environment in Ancient Life and Thought, Geographica Historica* 36, Stuttgart: Franz Steiner, pp. 11–24.
Coulson, F. T. (1991), *The Vulgate Commentary on Ovid's Metamorphoses: The Creation Myth and the Story of Orpheus*, Toronto: Pontifical Institute of Mediaeval Studies.
Coulson, F. (2007), 'Ovid's Transformations in Medieval France (ca. 1100 – ca. 1350)', in A. Keith and S. Rupp, eds, *Metamorphosis: The Changing Face of Ovid in Medieval and Early Modern Europe*, Toronto: Centre for Reformation and Renaissance Studies, pp. 33–60.
Coulson, F. (2011), 'Ovid's Metamorphoses in the school tradition of France, 1180–1400: Texts, manuscript traditions, manuscript settings', in J. T. Clark, F. T. Coulson and K. L. McKinley, eds, *Ovid in the Middle Ages*, Cambridge: Cambridge University Press, pp. 48–82.
Crapanzano, V. (1980), *Tuhami: Portrait of a Moroccan*, Chicago: University of Chicago Press.
Crocker, H. (2019), *The Matter of Virtue: Women's Ethical Action from Chaucer to Shakespeare*, Philadelphia: University of Pennsylvania Press.
Curley, D. (2013), *Tragedy in Ovid. Theater, Metatheater, and the Transformation of a Genre*, Cambridge: Cambridge University Press.
Curtius, E. R. (1953), *European Literature and the Latin Middle Ages*, trans. Willard R. Trask, London: Routledge.
Dan, A. (2011), 'Le Sang des Anciens: notes sur les paroles, les images et la science du sang', *Vita Latina* 183–184: 5–32.
Danowski, D., and E. Viveiros de Castro (2016), *The Ends of the World*, Malden, MA: Polity.
Davies, M. (2004), 'Aristotle Fr. 44 Rose: Midas and Silenus', *Mnemosyne* 57.6: 682–97.
Davis, P. J. (2006), *Ovid and Augustus: a Political Reading of Ovid's Erotic Poems*, London: Duckworth.
Dealy, R. (2017), *The Stoic Origins of Erasmus' Philosophy of Christ*, Toronto: University of Toronto Press.
de Boer, C., et al., eds (1915–38), *Ovide moralisé: poème du commencement du quatorzième siècle*, 15 (1915) (I, books 1–3); 21 (1920) (II, books 4–6); 30 (1931) (III, books 7–9); 37 (1936) (IV, books 10–13); 43 (1938) (V, books 14–15 and appendices), Amsterdam: Verhandelingen der Koninklijke Akademie van Wetenschapaen te Amsterdam, Afdeeling Letterkunde, 15.
Deleuze, Gilles, and Félix Guattari (1987), *A Thousand Plateaus: Capitalism and Schizophrenia*, trans. Brian Massumi, Minneapolis: University of Minnesota Press.
Della Corte, F. (1985), 'Gli *Empedoclea* ed Ovidio', *Maia* 37: 3–12.
DeLoughrey, E. (2019), *Allegories of the Anthropocene*, Durham, NC: Duke University Press.
Dench, E. (2005), *Romulus' Asylum. Roman Identities from the Age of Alexander to the Age of Hadrian*, Oxford: Oxford University Press.
Dent, R. W. (1964), 'Imagination in *A Midsummer Night's Dream*', *Shakespeare Quarterly* 15.2: 115–29.
Derrida, J. (1993), *Aporias*, trans. Thomas Dutoit, Stanford: Stanford University Press.
Derrida, J. (2008), *The Animal That Therefore I Am*, ed. Marie-Louise Mallet, trans. David Wills, New York: Fordham.
Descola, P. (2005), *Par delà nature et culture*, Paris: Gallimard.
Descola, P. (2013), *Beyond Nature and Culture*, trans. J. Lloyd, Princeton, NJ: Princeton University Press.

Desmond, M. (1989), 'Introduction', in *Ovid in Medieval Culture*, ed. Marilynn R. Desmond, *Mediaevalia* 13 (special issue): 1–7.
Detienne, M. (1972), *Les Jardins d'Adonis: La mythologie des aromates en Grèce ancienne*, Paris: Editions Gallimard. English translation (1994): *The Gardens of Adonis: Spices in Greek Mythology*, trans. Janet Lloyd, Princeton: Princeton University Press.
Döring, J. (1996), *Ovids Orpheus*, Frankfurt am Main: Stroemfeld.
Dronke, P. (1974), *Fabula: Explorations into the Uses of Myth in Medieval Platonism*, Leiden: Brill.
Duminil, M.-P. (1983), *Le Sang, les vaisseaux, le cœur dans la collection hippocratique. Anatomie et physiologie*, Paris: Les Belles Lettres.
Dwyer, K. (1982), *Moroccan Dialogues*, Baltimore: Johns Hopkins University Press.
Ebreo, L. (2009), *Dialogues of Love*, ed. R. Pescatori, trans. Damian Bacich and R. Pescatori, Toronto: University of Toronto.
Eco, U. (1975), *Trattato di semiotica generale*, Milan: Bompiani.
Emmett, R., and D. Nye (2017), *The Environmental Humanities: A Critical Introduction*, Cambridge, Mass: MIT Press.
Enterline, L. (2000), *The Rhetoric of the Body from Ovid to Shakespeare*, Cambridge: Cambridge University Press.
Erasmus (1549), *The Praise of Folie*, trans. Thomas Chaloner, London: Thomas Berthelet.
Evans, N. (2006), 'Diotima and Demeter as Mystagogues in Plato's *Symposium*', *Hypatia* 21.2: 1–27.
Fabre-Serris, J. (2018), 'Enjeux moraux et idéologiques des usages d'Empédocle au livre XV des *Métamorphoses*: une réponse d'Ovide à Virgile (*Enéide* VI et VIII),' in S. Franchet d'Espèrey and C. Lévi, eds, *Les Présocratiques à Rome*, Paris: Sorbonne Université Presses, pp. 303–19.
Fantham, E. (1979), 'Ovid's Ceyx and Alcyone: The Metamorphosis of a Myth,' *Phoenix* 33.4: 330–45.
Fantham, E. (2004), *Ovid's* Metamorphoses, Oxford: Oxford University Press.
Fahey, M. (2011), *Metaphor and Shakespearean Drama: Unchaste Signification*, London: Palgrave.
Feeney, D. (1999), '*Mea tempora*: Patterning of time in Ovid's *Metamorphoses*,' in P. Hardie, A. Barchiesi and S. Hinds, eds, *Ovidian Transformations: Essays on Ovid's Metamorphoses and its Reception*, Proceedings of the Cambridge Philological Society Supp. 23, Cambridge, pp. 13–30.
Feeney, D. (2004), 'Interpreting Sacrificial Ritual in Roman Poetry: Disciplines and their Models', in: A. Barchiesi, J. Rüpke and S. Stephens, eds, *Rituals in Ink: A Conference on Religion and Literary Production in Ancient Rome*, Stuttgart, pp. 1–21.
Feeney, D. (2007), *Caesar's Calendar: Ancient Time and the Beginnings of History*, Berkeley & Los Angeles: University of California Press.
Feldherr, A. (1997), 'Metamorphosis and Sacrifice in Ovid's Theban Narrative', *Materiali e discussioni per l'analisi dei testi classici* 38: 25–55.
Feldherr, A. (2002), 'Metamorphosis in the *Metamorphoses*', in P. Hardie, ed., *Cambridge Companion to Ovid*, Cambridge: Cambridge University Press, pp. 163–79.
Feldherr, A. (2010), *Playing Gods: Ovid's Metamorphoses and the Politics of Fiction*, Princeton/Oxford: Princeton University Press.
Finkelpearl, E., et al., eds (2014), *Apuleius and Africa*, New York: Routledge.
Flaig, E. (2007), 'Gladiatorial Games: Ritual and Political Consensus', *Journal of Roman Archaeology* 66: 83–92.
Fletcher, R. (2014), 'Prosthetic Origins: Apuleius the Afro-Platonist,' in E. Finkelpearl et al., eds, *Apuleius and Africa*, New York: Routledge, pp. 297–312.
Forey, M., ed. (2002), Ovid. *Metamorphoses*, trans. A. Golding, London: Penguin.
Formis, B., ed. (2009), *Penser en Corps. Soma-esthétique, art et philosophie*. Paris: L'Harmattan.
Formisano, M. (2021), 'Echo's revenge. Ovid's *Metamorphoses* and Spike Jonze's *her*', *Eugesta* 11, 2021: 221–47.

Forni, G. (2019), 'Semantica degli strumenti rurali in età romana Il caso dell'aratro: sua matrice ed evoluzione', in S. Segenni, ed., *L'agricoltura in età romana*, Consonanze 19, Milan: Ledizioni, pp. 157–203.
Francese, C. (2001), *Parthenius of Nicaea and Roman Poetry*, Studien Zur Klassischen Philologie 126, Frankfurt: Peter Lang.
Franklin-Brown, M. (2012), *Reading the World: Encyclopedic Writing in the Scholastic Age Bestiaries*, Chicago: University of Chicago Press.
Fratantuono, L. (2015), *A Reading of Lucretius'* De Rerum Natura, Lanham, Boulder, New York, London: Lexington Books.
Frontisi-Ducroux, F. (2017), *Arbres filles et garçons fleurs, Métamorphoses érotiques dans les mythes grecs*, Paris: Seuil.
Fuchs, T. (2017), 'Collective Body Memories', in C. Durt, T. Fuchs and C. Tewes, eds, *Embodiment, Enaction, and Culture. Investigating the Constitution of the Shared World*, Cambridge, Mass.: MIT Press, pp. 333–52.
Gale, M. (1994), *Myth and Poetry in Lucretius*, Cambridge: Cambridge University Press.
Garani, M. (2007), *Empedocles redivivus. Poetry and Analogy in Lucretius*, New York and Abingdon: Routledge.
Garrison, J. D. (1992), *Pietas from Virgil to Dryden*, Philadelphia: Pennsylvania State University Press.
Generosa, M. (1945), 'Apuleius and *A Midsummer Night's Dream*: Analogue or Source, Which?', *Studies in Philology* 42.2: 198–204.
Gibson, J. (1977), 'The Theory of Affordances', in R. Shaw and J. Bransford, eds, *Perceiving, Acting, and Knowing: Toward an Ecological Psychology*, Mahwah, NJ: Lawrence Erlbaum, pp. 67–82.
Gildenhard, I., and A. Zissos (1999), 'Problems of time in *Metamorphoses* 2', in P. Hardie, A. Barchiesi and S. Hinds, eds, *Ovidian Transformations: Essays on Ovid's Metamorphoses and its Reception*, Proceedings of the Cambridge Philological Society Supp. 23, Cambridge, 31–47.
Gildenhard, I. and A. Zissos (2013), 'The Transformations of Ovid's Medea', in *Transformative Change in Western Thought: A History of Metamorphosis from Homer to Hollywood*, Oxford: Legenda, pp. 88–130.
Giusti, E. (2016), 'Did Somebody Say Augustan Totalitarianism? Duncan Kennedy's "Reflections," Hannah Arendt's *Origins*, and the Continental Divide over Virgil's *Aeneid*', *Dictynna* 13.
Giusti, E. (2019), 'Bunte Barbaren Setting up the Stage. Re-Inventing the Barbarian on the *Georgics'* Theatre Temple (*G.* 3.1–48)', in B. Xinyue and N. Freer, eds, *Reflections and New Perspectives on Virgil's* Georgics, London: Bloomsbury, pp. 105–14.
Glare, P. G. W. (1996), *Oxford Latin Dictionary*, Oxford: Oxford University Press.
Glissant, É. (1997), *Poetics of Relation*, trans. B. Wing, Ann Arbor: The University of Michigan Press.
Glotfelty, C. and H. Fromm, eds, *The Ecocriticism Reader: Landmarks in Literary Ecology*, Athens, GE: University of Georgia Press.
Godfrey, P. and D. Torres, eds (2016), *Systemic Crises of Global Climate Change: Intersections of Race, Class, and Gender*, London/New York: Routledge.
Goldast, M., ed. (1610), *Ovidii Nasonis Pelignensis Erotica et amatoria opuscula*, Frankfurt: Wolfgang Richter.
Ghosh, A. (2016), *The Great Derangement*, Chicago: University of Chicago Press.
Goldsworthy, A. (2000), *Roman Warfare*, London: Cassell.
Gould, S. J. (1977), *Ontogeny and Phylogeny*, Cambridge, Mass.: Harvard University Press.
Gowers, E. (2005), 'Talking Trees: Philemon and Baucis Revisited', *Arethusa* 38: 331–65.
Gowers, E. (2011), 'Trees and Family Trees in the *Aeneid*', *Classical Antiquity* 30.1: 87–118.
Grady, H. (2008), 'Shakespeare and Impure Aesthetics: The Case of *A Midsummer Night's Dream*', *Shakespeare Quarterly* 59.3: 274–302.

Gragnolati, M., and F. Southerden (2020), *Possibilities of Lyric: Reading Petrarch in Dialogue*, Berlin: ICI Berlin Press.
Green, S. (2008), 'Save Our Cows? Augustan Discourse and Animal Sacrifice in Ovid's *Fasti*', *Greece & Rome* 55.1: 39–54.
Gregorić, P. (2005), 'Plato's and Aristotle's Explanation of Human Posture', *Rhizai: A Journal for Ancient Philosophy and Science* 2:183–96.
Griffin, A. H. F. (1991), 'Philemon and Baucis in Ovid's *Metamorphoses*,' *Greece & Rome* 38.1: 62–74.
Griffin, A. H. F. (1992), 'Ovid's Universal Flood,' *Hermathena* 152: 39–58.
Griffin, M. (2012), 'Translation and Transformation in the *Ovide moralisé*', in E. Campbell and R. Mills, eds, *Rethinking Medieval Translation*, Woodbridge: Boydell and Brewer.
Griffin, M. (2016), 'Imagining Ovid and Chrétien in Fourteenth-Century French Libraries', *French Studies* 70 (Special Issue: *The Medieval Library*): 201–15.
Guggenbühl, C. (1998), *Recherches sur la composition et la structure du ms Arsenal 3516*, Basel and Tübingen: A. Francke Verlag.
Guha, R., and J. M. Alier (1997), *Varieties of Environmentalism: Essays, North and South*, London: Earthscan Publications.
Guthrie, W. K. C. (1962), *A History of Greek Philosophy. Vol. 1. The Earlier Presocratics and the Pythagoreans*, Cambridge: Cambridge University Press.
Guthrie, W. K. C. (1993), *Orpheus and Greek Religion*, Princeton: Princeton University Press.
Habinek, T. (1990), 'Sacrifice, Society, and Virgil's Ox-born Bees', in M. Griffith and D. J. Mastronarde, eds, *Cabinet of the Muses: essays on classical and comparative literature in honor of Thomas G. Rosenmeyer*, Atlanta, pp. 209–23.
Hadot, P. (2006), *The Veil of Isis: An Essay on the History of the Idea of Nature*, trans. Michael Chase, Cambridge, Mass.: Harvard University Press.
Haeckel, E. (1866), *Generelle Morphologie der Organismen* (2 vols), Berlin: Reimer.
Hanson, V. D. (1998), *Warfare and Agriculture in Classical Greece*, Berkeley: University of California Press.
Haraway, D. J. (1997), *Modest_Witness@Second_Millennium.FemaleMan©_Meets_Onco MouseTM: Feminism and Technoscience*, London: Routledge.
Haraway, D. (2003), *The Companion Species Manifesto: Dogs, People, and Significant Otherness*. Paradigm 8. Chicago, IL: Prickly Paradigm Press.
Haraway, D. (2006), 'Encounters with Companion Species: Entangling Dogs, Baboons, Philosophers, and Biologists', *Configurations* 14.1–2: 97–114.
Haraway, D. (2008), *When Species Meet*, Minneapolis: University of Minnesota Press.
Haraway, D. (2016), *Staying with the Trouble: Making Kin in the Chthulucene*, Durham NC and London: Duke University Press.
Hardie, P. (1990), 'Ovid's Theban History: The First "Anti-Aeneid"?', *The Classical Quarterly* 40.1: 224–35.
Hardie, P. (1995), 'The Speech of Pythagoras in Ovid *Metamorphoses* 15: Empedoclean *Epos*', *The Classical Quarterly* 45.1: 204–14.
Hardie, P. (2002a), *Ovid's Poetics of Illusion*, Cambridge: Cambridge University Press.
Hardie, P. (2002b), 'Ovid and Early Imperial Literature,' in P. Hardie, ed., *The Cambridge Companion to Ovid*, Cambridge: Cambridge University Press, pp. 34–45.
Hardie, P., ed. (2015), *Ovidio: Metamorfosi, Volume VI: Libri XIII-XV*, trans. G. Chiarini, Milan: Mondadori.
Harf-Lancner, L., and M. Pérez-Simon (2015), 'Une lecture profane de l'*Ovide moralisé*. Le manuscrit BnF français 137 une mythologie illustrée', *Cahiers de recherches médiévales et humanistes* 30: 167–96.
Harper, K. (2019), *The Fate of Rome Climate, Disease, and the End of an Empire*, Princeton NJ: Princeton University Press.

Harrison, R. P. (1992), *Forests: The Shadows of the World*, Chicago and London: The University of Chicago Press.

Harrison, S. (2014), 'Ovid in Apuleius' *Metamorphoses*', in Miller and Newlands, eds, *A Handbook to the Reception of Ovid*, Oxford: Wiley-Blackwell, pp. 86–99.

Hatley, J. (2000), *Suffering Witness: The Quandary of Responsibility after the Irreparable*, New York: State University of New York Press.

Heckscher, W. S. (1956), 'The "Anadyomene" in the Medieval Tradition: Pelagia – Cleopatra – Venus,' *Nederlands Kunsthistorisch Jaarboek* 7: 1–38.

Heidegger, M. (1962), *Being and Time*, trans. John Macquarrie and Edward Robinson, San Francisco: Harper and Row.

Heise, U. (2016), *Imagining Extinction: The Cultural Meanings of Endangered Species*, Chicago, London: University of Chicago Press.

Hendricks, M. (1996), '"Obscured by Dreams": Race, Empire, and Shakespeare's *A Midsummer Night's Dream*', *Shakespeare Quarterly*, 47.1: 37–60.

Henkel, J. (2014), 'Virgil Talks Technique: Metapoetic Arboriculture in *Georgics* 2', *Virgilius* 60: 33–66.

Hexter, R., L. Pfuntner and J. Haynes (2020), *Appendix Ovidiana: Latin Poems Ascribed to Ovid in the Middle Ages*, Cambridge, Mass.: Harvard University Press.

Heyob, S. K. (1975), *The Cult of Isis among Women in the Graeco-Roman World*, Leiden: Brill.

Holinshed, R. (1587), 'Of Woods and Marishes,' in *Chronicles of England, Scotlande, and Irelande*, London, 211–14. [Open access transcript of 1577 and 1587 editions available from the Oxford Holinshed Project <http://english.nsms.ox.ac.uk/holinshed/texts.php?text1=1577_0074&text2=1587_0101#p1241>]

Holmes, B. (2010), *The Symptom and Subject: The Emergence of the Physical Body in Ancient Greece*, Princeton, NJ: Princeton University Press.

Holmes, B. (2016), 'The Manifest Life of Plants', in Caroline Picard, ed., *Imperceptibly and Slowly Opening*, Chicago: Green Lantern Press, pp. 68–73.

Holzberg, N. (1988), 'Ovids "Babylonika" (*Met.* 4.55–166)', *Wiener Studien*, 101: 265–77.

Hopkinson, N., ed. (2000), *Ovid: Metamorphoses Book XIII*, Cambridge: Cambridge University Press.

Horky, P. (2021), 'Pythaogrean Immortality of the Soul?,' in A. Long, ed., *Immortality in Ancient Philosophy and Theology: Life, Divinity and Identity*, Cambridge: Cambridge University Press, pp. 41–65.

Horsfall, N. (2001), *A Companion to the Study of Virgil*, 2nd ed. Leiden: Brill.

Hunt, A. (2010), 'Elegiac Grafting in Pomona's Orchard: Ovid, *Metamorphoses* 14.623–771', *Materiali e Discussioni per l'annalisi dei Testi Classici* 65: 43–58.

Hunt, A. (2016), *Reviving Roman Religion: Sacred Trees in the Roman World*, Cambridge: Cambridge University Press.

Hunt, A. (2019), 'Pagan Animism: A Modern Myth for a Green Age', in A. Hunt and H. Marlow, eds, *Ecology and Theology in the Ancient World: Cross-Disciplinary Perspectives*, London: Bloomsbury, pp. 137–52.

Huyck, J. (1987), 'Virgil's Phaethontiades', *Harvard Studies in Classical Philology* 91: 217–28.

Ingold, T. (1993), 'The Temporality of the Landscape', *World Archaeology* 25.2: 152–74.

Ingold, T. (2000), *The Perception of the Environment: Essays on Livelihood, Dwelling and Skill*, Abingdon: Routledge.

Ingold, T. (2006), 'Rethinking the Animate, Re-Animating Thought', *Ethnos: Journal of Anthropology* 71.1: 9–20.

Ingold, T. (2011), *Being Alive: Essays on Movement, Knowledge and Description*, Abingdon: Routledge.

Ingold, T. (2012), 'Toward an Ecology of Materials', *Annual Review of Anthropology* 41: 427–42.

Iovino, S. (2016), 'Posthumanism in Literature and Ecocriticism', *Relations* 4.1: 11–20.

Iovino, S., and S. Oppermann (2012), 'Material Ecocriticism: Materiality, Agency, and Models of Narrativity', *Ecozon@: European Journal of Literature, Culture and Environment* 3.1: 75–91.
Iovino, S., and S. Oppermann, S., eds (2014), *Material Ecocriticism*, Bloomington, IN: Indiana University Press.
Jacques, Z. (2013), 'Arboreal Myths: Transformation, Children's Literature, and Fantastic Trees', in I. Gildenhard and A. Zissos, eds, *Transformative Change in Western Thought*, London: Routledge, pp. 163–82.
Johnson, K., ed. (2009), *Virgil. The Georgics. A Poem of the Land*, trans. K. Johnson, London: Penguin.
Johnston, S. Iles (1997), 'Corinthian Medea and the Cult of Hera Akraia', in J. J. Clauss and S. Iles Johnston, eds, *Medea: Essays on Medea in Myth, Literature, Philosophy, and Art*, Princeton: Princeton University Press, 44–70.
Jones, O., and P. Cloke (2002), *Tree Cultures: The Place of Trees and Trees in their Place*. London: Berg.
Kahn, C. (2001), *Pythagoras and the Pythagoreans: A Brief History*, Indianapolis: Hackett Publishing.
Kay, S. (2017), *Animal Skins and the Reading Self in Medieval Latin and French Bestiaries*, Chicago: University of Chicago Press.
Keith, A. (2002), 'Sources and Genres in Ovid's Metamorphoses 1–5', in B. Weiden Boyd, ed., *Brill's Companion to Ovid*, Leiden and Boston: Brill, pp. 235–69.
Keith, A. (2017), 'Reception of Virgil's *Georgics* in Ovid's *Metamorphoses*', in P. Fedeli and G. Rosati, eds, *Ovidio 2017: prospettive per il terzo millennio: atti del Convegno Internazionale (Sulmona, 3/6 aprile 2017)*, Teramo: Ricerche & Redazioni, pp. 237–75.
Kennedy, D. F. (1992), '"Augustan" and "Anti-Augustan": Reflections on Terms of Reference,' in A. Powell, ed., *Roman Poetry and Propaganda in the Age of Augustus*, London, pp. 26–58.
Kenney, E. J., ed. (2011), *Ovidio. Metamorfosi*, vol. IV, Milan: Mondadori.
Kennedy, R. F., and M. Jones-Lewis (2016), *The Routledge Handbook of Identitiy and the Environment in the Classical and Medieval Worlds*, London: Routledge.
Kingsley, P. (1995), *Ancient Philosophy, Mystery, and Magic: Empedocles and the Pythagorean Tradition*, Oxford: Clarendon Press.
Kirksey, E., and S. Helmreich (2010), 'The Emergence of Multispecies Ethnography', *Cultural Anthropology* 25.4: 545–76.
Kohn, E. (2013), *How Forests Think: Toward an Anthropology Beyond the Human*, Berkeley: University of California Press.
Kleczkowska, K. (2017), 'Reincarnation in Empedocles of Akragas,' *Maska* 36: 183–98.
Klopsch, P. (1973), 'Carmen de philomela', in A. Önnerfors, J. Rathofer and F. Wagner, eds, *Literatur und Sprache im europäischen Mittelalter: Festschrift für Karl Langosch zum 70. Geburtstag*, Darmstadt: Wissenschaftliche Buchgesellschaft, pp. 173–94.
Kott, J. (1964), 'Titania and the Ass's Head', in *Shakespeare Our Contemporary*, trans. Boleslaw Taborski, Garden City, NY: Doubleday, pp. 207–28.
Kott, J. (1974), *Shakespeare Our Contemporary*, trans. B. Taborski, New York and London: Norton.
Kott, J. (1987), *The Bottom Translation*, trans. Daniela Miedzyrzecka and Lillian Vallee, Evanston: Northwestern University Press.
Krause, B. (2012), *The Great Animal Orchestra: Finding the Origins of Music in the World's Wild Places*, New York: Little, Brown.
Kristeva, J. (1984), *Revolution in Poetic Language*, trans. M. Waller, New York: Columbia University Press.
Kynes, W., ed. (2021), *Oxford Handbook of Wisdom and the Bible*, Oxford: Oxford University Press.
Lacoste-Dujardin, C. (1977), *Dialogue des femmes en ethnologie*, Paris: Maspero.

Lafaye, G. (1904), *Les Métamorphoses d'Ovide et Leurs Modèles Grecs*, Paris: Félix Alcan, Ancienne Librarie Germer Baillière.
LaFont, A., ed. (2014), *Shakespeare's Erotic Mythology and Ovidian Renaissance Culture*, Farnham: Ashgate.
Laks, A., and G. Most, eds (2016), *Early Greek Philosophy, Volume V: Western Greek Thinkers, Part 2*, Cambridge, Mass.: Harvard University Press (Loeb Classical Library).
Lamb, M. E. (1979), '*A Midsummer Night's Dream*: The Myth of Theseus and the Minotaur', *Texas Studies in Literature and Language* 21.4: 478–91.
Lamb, M. E. (2000), 'Taken by the Fairies: Fairy Practices and the Production of Popular Culture in *A Midsummer Night's Dream*', *Shakespeare Quarterly* 51.3: 277–312.
Lane, M. (2019), 'Ancient Ideas of Politics: Mediating between Ecology and Theology', in A. Hunt and H. Marlow, eds, *Ecology and Theology in the Ancient World: Cross-Disciplinary Perspectives*, London: Bloomsbury, pp. 13–23.
Langis, U. P. (2022), '"O Willow, Willow": *Othello*'s Desdemona as the Sufi-Sophianic Feminine,' forthcoming in K. Lehnhof, J. Reinhard Lupton, and C. Sale, eds, *Shakespeare's Virtuous Theater*, Edinburgh: Edinburgh University Press.
Langis, U. P., and J. Reinhard Lupton, eds (forthcoming), *Shakespeare and Wisdom Literature*, Edinburgh: Edinburgh University Press.
Largier, N. (2013), 'Figure, Plasticity, Affect,' in G. Brandstetter, G. Egert and S. Zubarik, eds, *Touching and Being Touched: Kinesthesia and Empathy in Dance and Movement*, Berlin and Boston: de Gruyter, pp. 23–34.
Laroque, F. (2014), 'Erotic Fancy/Fantasy in *Venus and Adonis*, *A Midsummer Night's Dream*, and *Antony and Cleopatra*,' in A. LaFont, ed., *Shakespeare's Erotic Mythology and Ovidian Renaissance Culture*, Farnham: Ashgate, pp. 61–76.
Latimer, J., and M. Miele, M. (2013), 'Naturecultures? Science, Affect and the Non-Human', *Theory, Culture & Society* 30 (7/8): 5–31.
Latour, B. (2013), *An Inquiry Into Modes of Existence*, trans. C. Porter, Cambridge, Mass.: Harvard University Press.
Latour, B. (2017), *Facing Gaia: Eight Lectures on the New Climatic Regime*, trans. C. Porter, Cambridge: Polity Press.
Latour, B. (2018), *Down to Earth: Politics in the New Climatic Regime*, trans. C. Porter, Cambridge: Polity Press.
Latour, B. (2021), *Ou suis-je? Leçons du confinement à l'usage des terrestres*, Paris: La Découverte.
Lejeune, P. (1975), *Le pacte autobiographique*, Paris: Le Seuil.
Lemaire, N. E., ed. (1824), *Poetae Latini Minores* (vol. 1), Paris: Didot.
Lennox, J. (2001), 'Are Aristotelian Species Eternal?', in *Aristotle's Philosophy of Biology*, Cambridge: Cambridge University Press, pp. 131–59.
LeVen, P. A. (2019), 'The Erogenous Ear', in Shane Butler and Sarah Nooter, eds, *Sound and the Ancient Senses*, London: Routledge, pp. 212–32.
LeVen, P. A. (2020), *Music and Metamorphosis in Graeco-Roman Thought*, Cambridge: Cambridge University Press.
Levinskaya, O. L., and M. Nikolsky (2017), 'Lucius or a New Io: On the Plot Shaping Technique in Apuleius' *Metamorphoses*', *Mnemosyne*, 4th Series, 70.1: 94–114.
Lévi-Strauss, C. (1962), *La Pensée sauvage*, Paris: Plon.
Lévi-Strauss, C. (1964), *Le Cru et le cuit*, Paris: Plon.
Lévi-Strauss, C. (1966), *Du Miel aux cendres*, Paris: Plon.
Lévi-Strauss, C. (1968), *L'Origine des manières de table*, Paris: Plon.
Lévi-Strauss, C. (1970), *The Raw and the Cooked*, New York: Harper & Row.
Lévi-Strauss, C. (1971), *L'Homme nu*, Paris: Plon.
Lévi-Strauss, C. (1971), *From Honey to Ashes*, New York: Harper & Row.

Lightfoot, J. L. (1999), *Parthenius of Nicaea: The Poetical Fragments and the Erōtika Pathēmata*, Oxford: Clarendon Press.
Lindley, D. (2005), *Shakespeare and Music*, London: Bloomsbury. (The Arden Shakespeare).
Linford, I. M. (1941), *The Arts of Orpheus*, Berkeley: University of California Press.
Littlefield, D. (1965), 'Pomona and Vertumnus: A Fruition of History in Ovid's *Metamorphoses*', *Arion* 4: 465–73.
Lloyd, G. E. R. (1968), *Aristotle: The Growth and Structure of his Thought*, Cambridge: Cambridge University Press.
Lloyd, W. (1699), *Life of Pythagoras*, London: H. Mortlock and J. Hartley.
Long, H. (1948), 'Plato's Doctrine of Metempsychosis and Its Source', *The Classical Weekly* 41.10: 149–55.
Lowe, D. (2010), 'The Symbolic Value of Grafting in Ancient Rome', *Transactions of the American Philological Association* 140: 461–88.
Luciani, S. (2018), 'Lucrèce et les psychologies présocratiques', in S. Franchet d'Espérey and C. Lévy, eds, *Les Présocratiques à Rome*, Presses Sorbonne Université.
Luciani, S., and M. L. Coletti (1981), 'Il problema del rapporto anima / sangue nella letteratura latina pagana', in *Atti della settimana Sangue e antropologia biblica*, Roma, pp. 331–47.
McColley, D. Kelsey (2007), *Poetry and Ecology in the Age of Milton and Marvell*, Oxford: Blackwell.
McCulloch, F. (1962), *Mediaeval Latin and French Bestiaries*, Chapel Hill, NC: University of North Carolina Press
McKibben, B. (1989), *The End of Nature*, New York: Random House.
McPeek, J. A. S. (1972), 'The Psyche Myth and *A Midsummer Night's Dream*', *Shakespeare Quarterly* 23.1: 69–79.
Mansfeld, J. (1987), 'Doxography and Dialectic. The Sitz im Leben of the Placita', in W. Haase, ed., *Philosophie, Wissenschaften, Technik. Philosophie*, Berlin: De Gruyter, pp. 3056–230.
Marcone, A. (2017), 'L'evoluzione della Sensibilità Ambientale a Roma all'inizio del Principato', in O. D. Cordovana and G. Chiai, eds, *Pollution and the Environment in Ancient Life and Thought*, Geographica Historica 36, Stuttgart: Franz Steiner, pp. 83–97.
Marcone, A. (2019), 'Agronomia e modelli di sviluppo a Roma tra la fine della Repubblica e l'Alto Impero', in S. Segenni, ed., *L'agricoltura in età romana*, Consonanze 19. Milan: Ledizioni, pp. 147–56.
Martelli, F. (2020), *Ovid. Research Perspectives on Classical Poetry 2.1,* Leiden: Brill.
Martin, C. (2005), *Ovid Metamorphoses. A New Translation*, New York: Norton.
Martin, P. M. (2009), *Res publica non restituta. La réponse d'Ovide: la légende de Cipus*, in F. Hurlet and B. Mineo, eds, *Le principat d'Auguste: Réalités et représentations du pouvoir. Autour de la Res publica restitut*, Rennes: Presses universitaires de Rennes.
Martinez, J. A. (1974), 'Galileo on Primary and Secondary Qualities,' *Journal of the History of the Behavioral Sciences* 10.2: 160–9.
Maschek, D. (2018), *Die römischen Bürgerkriege. Archäologie und Geschichte einer Krisenzeit*, Darmstadt: Philipp von Zabern.
Meisner, D. (2018), *Orphic Tradition and the Birth of the Gods*, Oxford: Oxford University Press.
Mencacci, F. (1986), '*Sanguis/cruor*. Designazioni linguistiche e classificazione antropologica del sangue nella cultura romana', *Materiali e discussioni per l'analisi dei testi classici* 17: 25–91.
Mentz, S. (2009), *At the Bottom of Shakespeare's Ocean*, London and New York: Continuum.
Mentz, S. (2021), *Ocean*, London: Bloomsbury.
Merkelbach, R., and M. L. West (1967), *Fragmenta Hesiodea*, Oxford: Clarendon Press.
Merrifield, A. (2009), *The Wisdom of Donkeys*, New York and London: Bloomsbury.
Mitchell, J. A. (2014), *Becoming Human: The Matter of the Medieval Child*, Minneapolis: University of Minnesota Press.
Monterese, F. (2012), *Lucretius and his Sources: A Study of* De Rerum Natura I, 635–920, Berlin and Boston: Walter de Gruyter.

Moore, J. W., ed. (2016), *Anthropocene or Capitalocene? Nature, History, and the Crisis of Capitalism*, Oakland, CA: PM Press.
Mortimer-Sandilands, C., and B. Erickson (2010), *Queer Ecologies: Sex, Nature, Politics, Desire*, Bloomington Indiana: Indiana University Press.
Morton, T. (2007), *Ecology without Nature: Rethinking Environmental Aesthetics*, Cambridge, Mass.: Harvard University Press.
Morton, T. (2010), *The Ecological Thought*, Cambridge, Mass.: Harvard University Press.
Morton, T. (2013), *Hyperobjects: Philosophy and Ecology after the End of the World*, Minneapolis: University of Minnesota Press.
Müller, H. (2001), *Macbeth, nach Shakespeare*. 1972. *Werke 4, Die Stücke 2* (ed. F. Hörnigk), Frankfurt am Main: Suhrkamp, 261–324.
Müller, H. (2002), *Der Auftrag – Erinnerung an eine Revolution*. 1979. *Werke 5, Die Stücke 3* (ed. F. Hörnigk), Frankfurt am Main: Suhrkamp, 11–42.
Müller, H. (2005), 'Shakespeare eine Differenz,' in *Werke 8, Schriften* (ed. F. Hörnigk), Frankfurt am Main: Suhrkamp, 334–7.
Myers, K. S. (1994), *Ovid's Causes: Cosmogony and Aetiology in the Metamorphoses*, Ann Arbor: University of Michigan Press.
Myers, N. (2015), 'Conversations on Plant Sensing: Notes from the Field', *NatureCulture* 3 (Acting with Non-Human Entities): 35–66.
Naess, A. (1953), *Interpretation and Preciseness: A Contribution to the Theory of Communication*, Oslo: Jacob Dybwab.
Naess, A. (1966), *Communication and Argument: Elements of Applied Semantics*, Oslo: Universitetsforlaget.
Naess, A. (1973), 'The Shallow and the Deep, Long-Range Ecology Movement,' *Inquiry* 16: 95–100.
Naess, A. (1985), 'The World of Concrete Contents,' *Inquiry* 28: 417–28.
Naess, A. (2008), 'An Example of a Place: Tvergastein', in A. Drengson, B. Devall, eds, *The Ecology of Wisdom. Writings by Arne Naess*, Berkeley: Counterpoint.
Nardizzi, V. (2013), *Wooden Os. Shakespeare's Theatres and England's Trees*, Toronto: University of Toronto Press.
Nardizzi, V. (2019), 'Daphne Described: Ovidian Poetry and Speculative Natural History in Gerard's *Herball*', *Philological Quarterly* 98.1–2: 137–56.
Nelis-Clément, J. (2017), 'Roman Spectacles: Exploring Their Environmental Implications', in O. D. Cordovana and G. F. Chiai, eds, *Pollution and the Environment in Ancient Life and Thought, Geographica Historica* 36, Stuttgart: Franz Steiner, pp. 217–81.
Nethercut, J. S. (2017), 'Empedocles' "Roots" in Lucretius' *De Rerum Natur*', *American Journal of Philology*, 138.1: 85–105.
Neuru, L. (1980), 'Metamorphosis: Some Aspects of This Motif in Ovid's *Metamorphoses*', PhD, Hamilton, Ontario: McMaster University.
Newlands, C. (1997), 'The Metamorphosis of Ovid's Medea', in J. J. Clauss and S. I. Johnston, eds, *Medea: Essays on Medea in Myth, Literature, Philosophy and Art*, Princeton, NJ: Princeton University Press, pp. 178–210.
Newmyer, S. T. (2014), 'Being the One and Becoming the Other: Animals in Ancient Philosophical Schools', *The Oxford Handbook of Animals in Classical Thought and Life*, Oxford: Oxford University Press, 507–34.
Nims, J., ed. (2000), *Ovid's Metamorphoses: The Arthur Golding Translation*, Philadephia: Paul Dry Books.
Nisbet, R. (1987), 'The Oak and the Axe: Symbolism in Seneca, *Hercules Oetaeus* 1618ff', in M. Whitby, P. Hardie and M. Whitby, eds, *Homo Viator: Classical Essays for John Bramble*, Bristol: Bristol Classical Press, pp. 244–51.
Nooter, Sarah (2019), 'Sounds of the Stage', in Shane Butler and Sarah Nooter, eds, *Sound and the Ancient Senses*, London: Routledge, pp. 198–211.

Nugent, G. (1989), 'This Sex Which Is Not One: De-Constructing Ovid's Hermaphrodite', *Differences* 2: 160–85.
Olson, S. D. (2012), *The Homeric Hymn to Aphrodite and Related Texts: Text, Translation and Commentary*, Berlin: De Gruyter.
Oppermann, S., and S. Iovino, eds (2017), *Environmental Humanities: Voices from the Anthropocene*, London: Rowman & Littlefield.
Ovid (1955), *Ovid: Metamorphoses*, trans. R. Humphries, Bloomington IN: Indiana University Press.
Ovid (2009), *Ovid. Metamorphoses,* trans. A. Melville, Oxford and New York: Oxford University Press.
Palmer, J. (2018), 'Presocratic Interest in the Soul's Persistence after Death', in J. E. Sisko, ed., *The History of the Philosophy of Mind* (Vol. 1): *Philosophy of Mind in Antiquity*, New York and London: Routledge, 23–43.
Parejko, K. (2003), 'Pliny the Elder's Silphium: First Recorded Species Extinction,' *Conservation Biology* 17.3: 925–7.
Paschalis, M. (2001), '*Semina Ignis*: The Interplay of Science and Myth in the Song of Silenus', *The American Journal of Philology* 122.2: 201–22.
Payne, M. (2010), *The Animal Part: Human and other Animals in the Poetic Imagination*, Chicago: Chicago University Press.
Payne, M. (2016), 'Trees in Shallow Time', in C. Picard, ed., *Imperceptibly and Slowly Opening,* Chicago: Green Lantern Press, pp. 76–83.
Peck, H. T. (1894), 'Onomatopoetic Words in Latin', in *Classical Studies in Honour of Henry Drisler*, New York: Macmillan, pp. 226–39.
Pellò, C. (2018), 'The Lives of Pythagoras: A Proposal for Reading Metempsychosis,' *Rhizomata* 6.2: 135–56.
Peirano, I. (2009), '*Mutati Artus*: Scylla, Philomela and the End of Silenus' Song in Virgil *Eclogue* 6', *The Classical Quarterly* 59.1: 187–95.
Perdue, L. (2007), *Wisdom Literature: A Theological History*, Louisville: Westminster Knox Press.
Perutelli, A. (1985), 'I 'bracchia' degli alberi: designazione tecnica e immagine poetica', *Materiali e Discussioni per l'annalisi dei Testi Classici* 15: 9–48.
Pettman, D. (2017), *Sonic Intimacy: Voice, Species, Technics (Or, How to Listen to the World)*, Palo Alto: Stanford University Press.
Pisanty, V. (2015), 'From the Model Reader to the Limits of Interpretation', *Semiotica* 206: 37–61.
Pitru, P. (2017), 'Life Form and Form of Life within an Agentive Configuration: A Birth Ritual among the Mixe of Oaxaca, Mexico', *Current Anthropology* 58.3: 360–80.
Plumwood, V. (1993), *Feminism and the Mastery of Nature*, New York and London: Routledge.
Possamaï-Pérez, M. (2006), *L'Ovide moralisé: essai d'interprétation*, Paris: Champion.
Powers, R. (2018), *The Overstory*, New York: W. W. Norton and Company.
Prescendi, F. (2007), *Décrire et comprendre le sacrifice: Les réflexions des Romains sur leur propre religion à partir de la littérature antiquaire*, Stuttgart: Franz Steiner Verlag.
Primavesi, O. (2007), 'Empedocle: divinité physique et mythe allégorique', *Philosophie antique* 7 (online since 13 May 2022); English translation as 'Empedocles: Physical and Mythical Divinity', in P. Curd and D. Graham, eds (2009), *Oxford Handbook of Presocratic Philosophy*, Oxford: Oxford University Press, pp. 250–83.
Race, W. H., ed. (2008), *Apollonius Rhodius: Argonautica*, trans. William H. Race, Cambridge, Mass.: Harvard University Press (Loeb Classical Library).
Radke-Stegh, M. (1978), *Der Theatervorhang. Ursprung – Geschichte – Funktion*, Meisenheim am Glan: Verlag Anton Hain.
Rehm, R. (2007), 'Festivals and audiences in Athens and Rome', in M. McDonald and M. Walton, eds, *The Cambridge Companion to Greek and Roman Theatre*, Cambridge: Cambridge University Press, pp. 184–201.

Reid, L. A. (2018), *Shakespeare's Ovid and the Spectre of the Medieval*, Cambridge: Boydell and Brewer.
Rhorer, C. C. (1980), 'Red and White in Ovid's *Metamorphoses:* The Mulberry Tree in the Tale of Pyramus and Thisbe', *Ramus* 9: 79–88.
Ribémont, B. (2002), 'L'*Ovide moralisé* et la tradition encyclopédique médiévale', *Cahiers de recherches médiévales et humanistes* 9, http://journals.openedition.org/crm/907.
Richter, S. (2015), *Pflug und Steuerruder. Zur Verflechtung von Herrschaft und Landwirtschaft in der Aufklärung*, Cologne: Böhlau.
Riedweg, R. (2002), *Pythagoras: Leben, Lehre, Nachwirkung: eine* Einführung, München: Beck. English translation (2011): *Pythagoras: His Life, Teaching, and Influence,* trans. S. Rendall, Ithaca: Cornell University Press.
Robertson, K. (2019), 'Scaling Nature: Microcosm and Macrocosm in Later Medieval Thought', *Journal of Medieval and Early Modern Studies* 49: 609–31.
Rose, D. B. (2012), 'Multispecies Knots of Ethical Time,' *Environmental Philosophy* 9: 127–40.
Roselaar, S. (2010), *Public Land in the Roman Republic: A Social and Economic History of Ager Publicus in Italy, 396–89 B.C.*, Oxford: Oxford University Press.
Rossetti, L. (2017), *Un altro Parmenide. Il sapere* peri physeos, *Parmenide e l'irrazionale* (Vol. 1), Bologna: Diogene.
Ruddiman, W. F. (2003), 'The Anthropogenic Greenhouse Era Began Thousands of Years Ago', *Climatic Change* 61: 261–93.
Rüpke, J. (2007), '*Religio* and *Religiones* in Roman Thinking', *Les Études Classiques* 75: 67–78.
Rust, M. D. (2007), *Imaginary Worlds in Medieval Books: Exploring the Manuscript Matrix*, New York: Palgrave Macmillan.
Ryan, M.-L. (1992), 'Possible Worlds in Recent Literary Theory', *Style* 26.4: 528–53.
Saudelli, L., and C. Lévy (2014), *Présocratiques latins. Héraclite*, Paris: Société d'Édition Les Belles Lettres.
Scaffai, N. (2017), *Letteratura e ecologia. Forme e temi di una relazione narrativa*, Rome: Carocci.
Scheid, J. (1998), 'L'Animal mis à mort: une interprétation romaine du sacrifice', *Études rurales* 147–8 (*Mort et mise à mort des animaux*, sous la direction de Anne-Marie Brisebarre): 15–26.
Scheid, J. (2011), 'Les offrandes végétales dans les rites sacrificiels des Romains', in V. Pirenne-Delforge et F. Prescendi, eds, *Kernos*, Supp. 26, Actes de la VIe rencontre du Groupe de recherche européen 'FIGURA'. *Nourrir les dieux? Sacrifice et représentation du divin dans les sociétés grecque et romaine*, Lièges, pp. 105–16.
Scheid, J. (2012), 'Roman animal sacrifice and the system of being', in F. Naiden and C. Faraone, eds, *Greek and Roman Animal Sacrifice: Ancient Victims, Modern Observers*, Cambridge: Cambridge University Press, pp. 84–96.
Scheid, J., and J. Svenbro (1996), *The Craft of Zeus: Myths of Weaving and Fabric,* trans. C. Volk, Cambridge, Mass.: Harvard University Press.
Schiesaro, A. (2003), *The Passions in Play: Thyestes and the Dynamics of Senecan Drama*, Cambridge University Press.
Schilhab, T. (2017), 'Embodied Cognition and Science Criticism: Juxtaposing the Early Nietzsche and Ingold's Anthropology', *Biosemiotics* 10.3: 469–76.
Schilpp, A. (1949), *Albert Einstein: Philosopher-Scientist,* Evanston, IL: Evanston University Press.
Schimmel, A. (1992), *A Two-Colored Brocade: The Imagery of Persian Poetry*, University of North Carolina Press.
Schimmel, A. (1997), *My Soul Is a Woman: The Feminine in Islam,* trans. Susan H. Ray, New York: Continuum.
Schimmel, A. (2001), *As Through a Veil: Mystical Poetry in Islam*, Oxford: Oneworld.
Schliephake, C., ed. (2016), *Ecocriticism, Ecology, and the Cultures of Antiquity*, Lanham, MD: Lexington Books.

Schliephake, C. (2020), *The Environmental Humanities and the Ancient World*, Cambridge: Cambridge University Press.
Schmitzer, U. (1990), *Zeitgeschichte in Ovids Metamorphosen. Mythologische Dichtung unter politischem Anspruch*, Stuttgart: Teubner.
Scott, C. (2014), *Shakespeare's Nature: From Cultivation to Culture*, Oxford: Oxford University Press.
Sedley, D. (1998), *Lucretius and the Transformation of Greek Wisdom*, Cambridge: Cambridge University Press.
Segal, C. (1969a), 'Virgil's Sixth *Eclogue* and the Problem of Evil', *Transactions and Proceedings of the American Philological Association* 100: 407–35.
Segal, C. (1969b), 'Myth and Philosophy in the *Metamorphoses*: Ovid's Augustanism and the Augustan Conclusion of Book XV', *The American Journal of Philology* 90: 257–92.
Segal, C. (1989), *Orpheus: The Myth of the Poet*, Baltimore: Johns Hopkins University Press.
Segal, C. (2001), 'Intertextuality and Immortality: Ovid, Pythagoras and Lucretius in *Metamorphoses* 15', *Materiali e discussioni per l'analisi dei testi classici* 46: 63–101.
Seider, A. (2016), 'Genre, Gallus, and Goats: Expanding the Limits of Pastoral in *Eclogues* 6 and 10', *Virgilius* 6: 3–23.
Setaioli, A. (1998), 'L'impostazione letteraria del discorso di Pitagora nel XV libro delle *Metamorfosi*,' in H. von Werner Schubert, ed., *Ovid Werk und Wirkung*, Bern: Peter Lang, 487–514.
Sharrock, A. (1994), 'Ovid and the Politics of Reading', *Materiali e Discussioni per l'analisi Dei Testi Classici*, 33: 97–122.
Sharrock, A. (1996), 'Representing Metamorphosis', in J. Elsner, ed., *Art and Text in Roman Culture*, Cambridge: Cambridge University Press, pp. 103–30.
Sherman, D. (2016), *Second Death: Theatricalities of the Soul in Shakespearean Drama*, Edinburgh: Edinburgh University Press.
Shusterman, R., ed. (2018), *Aesthetic Experience and Somaesthetics*, Leiden and Boston: Brill.
Sillars, S. (2013), 'Parody and the Erotic Beast: Relocating Titania and Bottom', in Agnés LaFont, ed., *Shakespeare's Erotic Mythology and Ovidian Renaissance Culture*, Farnham: Ashgate, pp. 107–23.
Simmonds, P. M. (1992), *Myth, Emblem and Music in Shakespeare's* Cymbeline: *An Iconographic Reconstruction*, Newark DE: University of Delaware Press.
Sissa, G. (2010), 'AMOR MORA METAMORPHOSIS ROMA', in M. De Poli, ed., *Maschile e Femminile: Genere ed Eros nel Mondo Greco* (Atti del Convegno, Università degli Studi di Padova, 22–23 ottobre 2009), Padova, pp. 7–38.
Sissa, G. (2019), 'Apples and Poplars, Nuts and Bulls: The poetic biosphere of Ovid's *Metamorphoses*', in Emanuela Bianchi et al., eds, *Antiquities Beyond Humanism*, Oxford: Oxford University Press, pp. 159–86.
Skutsch, O. (1956), 'Zu Virgils Eklogen', *Rheinisches Museum Für Philologie* 99: 13–201.
Smith, A. (1997), *Poetic Allusion and Poetic Embrace in Virgil and Ovid*, Ann Arbor, MI: The University of Michigan Press.
Solodow, J. B. (1988), *The World of Ovid's* Metamorphoses, Chapel Hill and London: University of North Carolina Press.
Souilhé, J. (1932), 'L'Énigme d'Empédocle', *Archives de Philosophie* 9.3 (*Études d'Histoire de la Philosophie*): 1–23.
Spencer, D. (2019), *Language and Authority in* De Lingua Latina: *Varro's Guide to Being Roman*, Madison, WI: The University of Wisconsin Press.
Stanford Encyclopedia of Philosophy, last updated May 1, 2018, https://plato.stanford.edu/entries/locke/
Starks, L., ed. (2019), *Ovid and Adaptation in Early Modern Theatre*, Edinburgh: Edinburgh University Press.
Starks-Estes, L. S. (2014), *Violence, Trauma, and Virtus in Shakespeare's Roman Poems and Plays: Transforming Ovid*, New York: Palgrave Macmillan.

Starnes, D. T. (1945), 'Shakespeare and Apuleius', *Publications of the Modern Language Association of America* 60.4: 1021–50.
Staton, W. F. (1963), 'Ovidian Elements in *A Midsummer Night's Dream*', *Huntington Library Quarterly* 26.2: 165–78.
Stein, C. (2013), 'Beyond the Generation of Leaves: The Imagery of Trees and Human Life in Homer', for unpublished diss. UCLA.
Tallis, R. (2017), *Of Time and Lamentation: Reflections on Transience*, Newcastle upon Tyne: Agenda.
Tarrant, R., ed. (2004), *P. Ovidius Nasonis Metamorphoses*, Oxford: Oxford University Press.
Tatum, W. J. (1984), 'The Presocratics in Book One of Lucretius' *De Rerum Natura*', *Transactions of the American Philological Association* 114: 177–89.
Taylor, A. B., ed. (2000), *Shakespeare's Ovid: The Metamorphoses in the Plays and Poems*, Cambridge: Cambridge University Press.
Taylor, C. C. W. (1999), *The Atomists: Leucippus and Democritus. Fragments, A Text and Translation with Commentary,* Toronto: University of Toronto Press.
Taylor, G., et al., eds (2016), *New Oxford Shakespeare: Modern Critical Edition*, Oxford: Oxford University Press.
Thompson, D. W. (1894), *A Glossary of Greek Birds*, Oxford: Clarendon Press.
Thorne, J. A. (2007), 'Warfare and Agriculture: The Economic Impact of Devastation in Classical Greece', in E. L. Wheeler, ed., *The Armies of Classical Greece*, Aldershot: Ashgate, pp. 195–224.
Trippett, D. (2018), 'Music and the Transhuman Ear: Ultrasonics, Material Bodies, and the Limits of Sensation', *The Musical Quarterly*, 100.2: 199–261.
Tsing, A. (2017), *The Mushroom at the End of the World: On the Possibility of Life in Capitalist Ruins*, Princeton: Princeton University Press.
Tsitsiou-Chelidoni, C. (2003), *Ovid, Metamorphosen Buch VIII: narrative Technik und literarischer Kontext*, Frankfurt am Main: Peter Lang.
Uexküll, J. von (1956), *Streifzuge durch die Umwelten von Tieren und Menschen: ein Bilderbuch unsichtbarer Welten, Bedeutungslehre*, Hamburg: Rowohlt.
Uexküll, J. von (1957), 'A Stroll through the World of Animals and Men: A Picture Book of Invisible Worlds', in C. H. Schiller, ed. and trans., *Instinctive Behavior: The Development of a Modern Concept*, New York: International Universities Press, pp. 5–80.
Uzgalis, W. *Stanford Encyclopedia of Philosophy*, s.v. 'John Locke,' last updated 1 May 2018. https://plato.stanford.edu/entries/locke/
Van Dooren, T. (2014), *Flight Ways: Life and Loss at the Edge of Extinction*, New York: Columbia University Press.
Vial, H. (2013), 'Le(s) César(s) d'Ovide', in O. Devillers and G. Flamerie de Lachapelle, eds, *Poésie augustéenne et Mémoires du passé de Rome*, Bordeaux: Éditions Ausonius, pp. 127–39.
Vico, G. (2020), *The New Science,* trans. J. Taylor and R. Miner, New Haven, CT and London: Yale University Press.
Waldenfels, B. (2002), *Bruchlinien der Erfahrung. Phänomenologie – Psychoanalyse – Phänomenotechnik*, Frankfurt am Main: Suhrkamp.
Waldenfels, B. (2010), *Sinne und Künste im Wechselspiel. Modi ästhetischer Erfahrung*, Frankfurt am Main: Suhrkamp.
Waldron, J. (2012), 'Phenomenology and Theater: "The Eye of Man Hath Not Heard": Shakespeare, Synesthesia, and Post-Reformation Phenomenology', *Criticism* 54.3: 403–17.
Walton, J. M., ed. (1988), Euripides: *Plays 1: Medea – The Phoenician Women – Bacchae,* trans. J. M. Walton, London: Methuen.
Wasdin, K. (2017), 'Weaving Time: Ariadne and the Argo in Catullus, C. 64', *Helios* 44.2: 181–99.
Watkins, C. (2011), *The American Heritage Dictionary of Indo-European Roots*, 3rd ed., Boston: Houghton Mifflin Harcourt.
Watson, R. (2006), *Back to Nature: The Green and the Real in the Late Renaissance*, Philadelphia PA: University of Pennsylvania Press.

Weber, M. (1891), *Die römische Agrargeschichte in ihrer Bedeutung für das Staats- und Privatrecht*, Stuttgart: Ferdinand Enke.
Weeda, L. (2015), *Virgil's Political Commentary in the* Eclogues, Georgics *and* Aeneid, Berlin and Boston: de Gruyter.
Weinstock, S. (1960), 'Pax and the "Ara Pacis"', *Journal of Roman Studies* 50.1–2: 44–58.
Welch, K. E. (2007), *The Roman Amphitheatre: From Its Origins to the Colosseum*, Cambridge: Cambridge University Press.
West, M. L. (1983), *The Orphic Poems*, Oxford: Clarendon Press.
Wheeler, S. (1995) '*Imago Mundi*: Another View of the Creation in Ovid's *Metamorphoses*', *The American Journal of Philology* 116.1: 95–121.
Wheeler, S. (2000), *Narrative Dynamics in Ovid's Metamorphoses*, Tübingen: Classica Monacensia.
Wind, E. (1968), *Pagan Mysteries in the Renaissance*, New York: W. W. Norton.
Wohlleben, P. (2016), *The Hidden Life of Trees: What They Feel, How They Communicate – Discoveries from a Secret World*, Vancouver: Greystone Books.
Wolfe, C. (2003), 'Subject to Sacrifice: Ideology, Psychoanalysis, and the Discourse of Species in Jonathan Demme's *The Silence of the Lambs*', Chapter 3 in *Animal Rites*, Chicago: Chicago University Press, 97–121.
Wood, D. (2005), *The Step Back: Ethics and Politics after Deconstruction*, Albany, NY: SUNY Press.
Wreford, A., D. Moran and N. Adger (2010), *Climate Change and Agriculture: Impacts, Adaptation and Mitigation*, Paris: OECD Publishing.
Wright, M. R., ed. (1995), *Empedocles: The Extant Fragments*, London: Yale University Press.
Zatta, C. (2016), 'Plants' Interconnected Lives: from Ovid's Myths to Presocratic Thought and Beyond', *Arion* 24.2: 101–26.
Zatta, C. (2018), 'Aristotle on the Sweat of the Earth', *Philosophia* 48: 55–70.
Zatta, C. (2020), 'La plasticité du corps et l'intériorité de plantes dans les mythes d'Ovide et au-delà', in M. W. De Bono, ed., *L'intelligence des plantes en question*, Paris: Hermann, pp. 143–57.
Zeitlin, F. I. (1990), 'Thebes: Theater of Self and Society in Athenian Drama', in J. Winkler and F. I. Zeitlin, eds, *Nothing to Do with Dionysos? Athenian Drama in Its Social Context*, Princeton: Princeton University Press, pp. 130–67.

INDEX

Index

Milton Keynes UK
Ingram Content Group UK Ltd.
UKHW020407170823
427012UK00003B/154